The Algebra

of

Data

A Foundation for the Data Economy

by Professor Gary Sherman
& Robin Bloor, Ph.D.

THE BLOOR GROUP PRESS

THE BLOOR GROUP PRESS

Austin, Texas

THE ALGEBRA OF DATA

First Edition: June 2015

Second Edition: December 2016

Third Edition: January 2018

Library of Congress Control Number: 2015950235

ISBN 978-0-9789791-6-4

Printed in the United States of America

Dedication

To Georg Cantor, the father of set theory and the grandfather of data algebra.

An invasion of armies can be resisted, but not an idea whose time has come.

~Victor Hugo

The Algebra

of

Data

Introduction

We stand at the dawn of the data economy. Fundamental changes have been set in motion that will affect the lives of people, industries and governments in dramatic ways. A powerful confluence of exciting technology developments and business innovations is driving it.

The data economy was heralded by big data, which acquired a life of its own, engendering a host of new software and enabling businesses to amass and analyze large collections of data. Big data likely accelerated the sudden emergence of the Internet of Things, soon to be populated by billions of sensors and connected devices generating more data than anyone could have ever imagined.

The big cloud players – Amazon, Microsoft, Google, and others – began to transforming computing into a utility delivering benefit consumers and enterprises far and wide.

Then, suddenly emerging from left field, blockchain technology introduced a whole new domain of distributed computing, organizing data in completely new ways and providing genuine digital currencies, governed by consensus and completely secure. IT introduced the wholly new idea of automatically enforced smart contracts and provably trusted networks. Another revolution, perhaps as far reaching as the Internet has begun.

All of this is underpinned by the ceaseless march of technology, as almost every aspect of computers and networks continues to evolve. Billions of computers and the high-speed networks that bind them together are the bedrock of the data economy.

So what is the data economy?

In my view, it is a commercial environment enhanced and governed by the mathematical analysis of data in every form, from real-time streams to the vast collections of stored data. In this emerging economy, enterprises will be guided by the knowledge and projections that they can tease out from their own data, combined with the growing amounts of external and public data that now abounds. Not far in the future, organizations will live or die by their ability to manage and analyze data.

They will use it to better understand their customers, to improve their decisions, to automate and leverage real-time business processes and to create innovative products and services – in short, to be competitive. And it's not just commercial businesses that will become stronger, smarter and more efficient. So will all areas of government and the non-profit sector.

It is my belief and expectation that the algebra of data will be a key foundation for the data economy. It will enable and facilitate the integration of the vast oceans of data that are the raw material of the data economy, and it will help optimize the retrieval, manipulation and management of all this data.

I am not a mathematician. So to a degree I played the role of a bystander who managed Algebraix Data Corporation and organized the funding that facilitated the development team's extraordinary effort in devising the algebra of data.

What kept me going during these intense years? Along the way, I discussed what we were doing with scores of computer science experts and mathematicians, including Robin Bloor, co-author of this book. Without exception, they validated the premise and promise of data algebra. Like me, they believed that once it was tested and proven, data algebra would provide a new foundation for an extensive range of computer applications.

Now, as I witness the growing impact of the data economy, I believe even more strongly that data algebra will play a big role in paving the way for ever more sophisticated uses of data, innovations that will benefit the world in ways we can't even guess at today.

With the publication of this book – undoubtedly the first of many that will be written on this topic – data algebra is being made available to anyone and everyone with some mathematical skills, a reasonable knowledge of software and a belief in the future's possibilities. I encourage you to read about data algebra, to gather an understanding of it and to apply it.

Charles Silver, CEO, Algebraix Data Corporation

Advice to the Reader

Mathematics is a more powerful instrument of knowledge than any other that has been bequeathed to us by human agency.

~ Descartes

⎯⎯∞⎯⎯

The purpose of this book is to describe and explain data algebra in a way that makes it as accessible as possible to the reader. While the book serves only as an introduction to data algebra, it is enough of an introduction to reveal the nature of data algebra and its applicability to software. As such, it paves the way for a further book that goes "broader and deeper" and will resemble a mathematics textbook more than this one does.

We expect some readers to purchase this book entirely out of curiosity – to see what this algebra is. To cater to such readers, we split the book between chapters that explain and discuss the mathematics and others that talk around it. If you are such a reader, you will probably be most interested in the "not-so-mathematical" chapters of the book: Chapters 1, 2, 5 and 9.

For the serious reader whose goal is to absorb and understand the mathematics, we have used every trick we could think of to make the subject as approachable as a mathematics book is ever likely to be. Where possible, we have employed a whimsical writing style rather than adhering to the dry mode of exposition that one usually encounters in mathematics books. We have also done our best to make the examples as entertaining as possible.

Nevertheless, this is a book about mathematics. Data algebra is not difficult to understand, but it is not a cake walk either. The subject matter is challenging, and unless you are a natural mathematician, you are best advised to absorb the mathematical chapters one concept at a time and make notes as you go. If you take your time, you will have a good grasp of data algebra by the time you arrive at the last page of the book.

Here's the rub for the novice. Data algebra is about things that are simultaneously very general, and therefore intimidating to some, while at the same time elementary in the formal mathematical sense. Entry level number theory is an example of a topic that is considered elementary while being surreptitiously difficult, mainly because of the sophisticated mathematics to which it eventually leads.

So it is with data algebra. The algebra of data is abstract and general, which is a wee bit scary. Nevertheless, it sprouts immediately from self-evident assumptions – that is, axioms of ZFC set theory – and it does not get any more "elementary" than that.

If you are a software developer then you might like to pursue the subject further by downloading and experimenting with the Python libraries that are described in Appendix B.

Acknowledgements

The authors acknowledge the assistance of Algebraix Data Corporation, who provided access to relevant staff to discuss the application of data algebra to software. We are particularly indebted to Charles Silver, CEO of Algebraix Data Corporation for his support in this activity, and also to the following individuals, all of whom helped in the production of this book in one way or another: Val Weaver, Rich Benci, Zoiner Tejada, Wes Holler, Gerhard Fiedler, Rebecca Jozwiak and Judy Ryser.

Table of Contents

—◦◦◦—

Chapter 1: What Is?

There is nothing more difficult to take in hand, more perilous to conduct, or more uncertain in its success, than to take the lead in the introduction of a new order of things.

~ Machiavelli

———— ❈ ————

WE TOOK Machiavelli's advice to heart. The algebra of data is a new application of mathematics. It has existed as a set of mature usable techniques since 2012. The few experienced mathematicians who are currently familiar with it will tell you there is nothing new about it, in the pure mathematical sense. For example, there was no need to adjust or extend set theory to formulate the various algebras that constitute the algebra of data – yes, indeed, data algebra comprises more than one algebra.

It applies set theory in a new way, and for that reason, it constitutes a new branch of applied mathematics. It was first formulated[1] as a collection of mathematical constructs that were thought to provide a practical foundation for data management software. It was gradually tested and applied to data management problems, first to "relational" data structures and then to "graph" data structures.

Algebraix Data Corporation funded its invention and development with the goal of creating mathematically based software products and services. In 2013, the company published a paper explicitly describing the new algebra, but it was in a form intended only for experienced mathematicians, and it generated little interest. The paper is called *Data Algebra, Hiding in Plain Sight* and can be downloaded at www.algebraixdata.com.[2]

While the paper provides a purely mathematical reference, the aim of this book is distinctly different. It intends to make readers familiar enough with data algebra to be able to use it. To that end, a small team of developers has created an "Algebraix Library" (initially written in Python). It is available

---

1   The formulation and development of the algebras that make up the algebra of data was carried out primarily by Professor Gary Sherman, one of the authors of this book.
2   The URL is: http://www.algebraixdata.com/wp-content/uploads/2013/03/DAHIPS-Web-ver-1.1.pdf.

free to all readers with programming skills who wish to experiment with data algebra.

This initial chapter is devoted to answering several questions that all begin with the words "What is..." We believe they are worth discussing before we introduce the algebra of data.

The first such question is:

What is mathematics?

The *New Oxford American Dictionary* defines it as:

> The abstract science of number, quantity, and space. Mathematics may be studied in its own right (pure mathematics), or as it is applied to other disciplines such as physics and engineering (applied mathematics).

Alternatively, the *Merriam-Webster* dictionary says:

> The science of numbers and their operations, interrelations, combinations, generalizations, and abstractions and of space configurations and their structure, measurement, transformations, and generalizations.

In fact, neither of these definitions is entirely adequate. The first confines mathematics to number, quantity and space, when in practice, some areas of mathematics deal with "things" that may not be spatial or numeric or have the attribute of quantity. Data can be such a thing.

The second definition is slightly better, as it refers directly to operations, interrelations, combinations, generalizations and abstractions, which are the beating heart of mathematics. Nevertheless, this definition also confines mathematics to numeric and spatial concerns.

The power of mathematics lies partly in its ability to construct useful models of real things. Using scale measurements, geometry and various calculations, you can design buildings. With more sophisticated mathematics, you can design chemical plants or airplanes. Mathematics is both the toolbox of innovators and the workhorse of engineers.

Even where mathematics has failed to model a domain perfectly – for example, weather prediction – it has no peer. Weather forecasts still rely on mathematical models, and nowadays, given sufficient computer power and historical data, they are usually accurate for two days or more.

In fact – and this is important to understand – mathematics is never wrong per se. As Isaac Newton remarked, "Errors are not in the art but in the artificers." A mathematical statement can be incorrect if someone makes an error in formulating it. A mathematical model can be horribly inaccurate if the modeling work is done poorly. But the mathematics itself is blameless. The tool is either used well or badly. This can be said of very few areas of study.

Data algebra models data, and it does so with rigorous accuracy. This attention to detail makes it an extremely useful tool for anyone who wants to define data structures and manipulate them algebraically.

The next question is also a fundamental one:

What is a computer?

Dictionaries focus on the fact that a computer is an electronic machine that runs programs to process information. However, here we need to be more particular and define a computer in terms that have relevance to data algebra.

So think of a computer as an electronic machine that runs programs that *act on data*. This brief, accurate definition allows us to make an important point:

> *The only thing a computer ever does is take data and transform it into other data.*

Every program that you have ever used or can conceive of does this and only this. It is true of operating systems, spreadsheets, email and big databases. It is true of every executable program.

There is an extremely powerful corollary to this: given that a data algebra exists, ALL programs can be written in algebraic terms as a set of mathematical functions that transform various algebraic collections of data into other algebraic collections of data.

Whether this means that ALL programs should be written in a strict algebraic manner is a moot point. In reality, simple programs will gain little from it.

What is an algebra?

Back to the dictionaries. *Merriam-Webster's* answer is:

> Any of various systems or branches of mathematics or logic concerned with the properties and relationships of abstract entities (as complex numbers, matrices, sets, vectors, groups, rings, or fields) manipulated in symbolic form under operations often analogous to those of arithmetic...

This is quite an accurate description of algebra. Algebraic statements use symbols (+, −, ×, ÷, ±, =, ≠, ∈, ∪, ∧ and many others) as a shorthand to express specific relationships quickly and succinctly. Such symbols form a kind of lexicon that mathematicians quickly learn. But some people find these symbols difficult to follow and are turned off, perhaps unnecessarily, from mathematics.

To minimize this problem for readers, we have gone to some lengths to translate algebraic statements into words and provided illustrations as well. Of course, those who quickly pick up the symbols can skip these aids.

There are many algebras.

Wikipedia provides a list of over fifty distinct algebras, each focusing on different targets – and it is not an exhaustive list. Data algebra itself is not a single algebra, as we have already noted. You will appreciate what this means as you progress through the book. You will realize that it is, in reality, a number of algebras that are included, one within another, like Russian nesting dolls.

Algebras are specifically built to enable abstraction and generalization. In algebra, the use of alphabetic letters or other symbols to represent real things is a fundamental tactic. So when we say, for example, a locomotive is travelling due West at 60 mph, we may refer to the locomotive as, say, L and its velocity as *v*. Thus we abstract a real-world object by using a single letter and its velocity by using another letter.

If we manage through various manipulations to find relationships between some of these letters, regardless of their value in any given context, we have a generalization that can be applied to many contexts, which may be very useful. One simple example is the equation you probably learned at school:

$$v(t) = v_0 + at,$$

where v = velocity, v_0 = initial velocity, a = acceleration and t = time.

Discovering such relationships is the primary purpose of any given algebra when it is used to model some real world situation. However, with data algebra the situation is slightly different. The goal of the algebra is to enable data to be handled algebraically. It does not model the data, it defines it. The rewards that the algebra naturally bestows are generalizations that can be applied to all data. It can solve problems in data integration and software optimization that might otherwise be intractable.

What is data?

In order to construct a data algebra, it is necessary to find an answer to the question "What is data?" and, having found an answer, to discover productive ways to represent data algebraically. Let's satisfy your immediate curiosity – assuming you have some – by discussing the question.

So, if you are willing to play, you are up. Think about the question "What is data?" for a moment or two, and then write your answer in pencil in the space below, without peeking at the text that follows:

-

-

Dictionary definitions of data make assertions of the following kind:

- Facts and statistics collected together for reference or analysis.
- Factual information used as a basis for reasoning.
- Any representations to which meaning is or might be assigned.
- The plural of datum.

Thesauri suggest confusing synonyms including: information, aggregation, collection, accumulation, assemblage, fact, details and even input. Perhaps this is what we should expect from the compilers of dictionaries and thesauri, but it's a little disappointing, and it doesn't fry any fish.

The responses to this question from IT experts in the field, often preceded by a long pause and punctuated with some facial contortions and guttural utterings, are more telling. We have experienced the following replies:

- "Knowledge."
- "Facts."
- "Files, digital information."
- "A database."
- "Anything that somebody might want to know that is enumerated, listed or quantified."
- "It is the answer to a question (may be a hypothetical question)."
- "It can be written (this includes that it is limited)."
- "It's data. It is what it is."
- "In computer parlance, it literally means anything that has been captured in any form on any digital recording medium."

We have even met with the somewhat defensive, "What do you mean by that?" and the somewhat facetious, "A character on Star Trek: Next Generation."

This sample of responses is a hodgepodge of ambiguity and circularity which, if left undisturbed, precludes a rigorous mathematical discussion of data, and since that's what we're focused on, we must do better. If we do not, we will simply repeat the previous errors of the IT industry, which, for its sins, gave birth to SQL, a mathematical disaster of the first order.

To paraphrase Carl Sagan (*Dragons of Eden*), those closest to the intricacies of data seem to have a more highly developed (and ultimately erroneous) sense of its mathematical intractability than those at some remove.

As it turns out, the wife of one of the authors, a notorious data-phobe, provided the most suggestive responses to both questions.

"What is data?" was met with a quizzical look, then, "Well, come on now, you and I both know what it means."

And the subsequent question "What is a datum?" was met with a more menacing look, and "You and I both know what it means!"

Indeed, anybody making a living in the data world knows what data is, and knows what a datum is, and uses the terms with reckless abandon in discussions with colleagues who also know what data is and know what a datum is, even though such "undoubted knowledge" calls into question the very definition of the word "know."

The only implied constant in such discussions is suggested by the surprisingly useful dictionary definition: "data is the plural of datum." That is, data (whatever it is) is a collection of "datums," and a datum (whatever that is) may be collected with other "datums" to create data.

Our assertion here is that the key to providing a practical mathematical understanding of data is to take data and datum as primitive undefined terms and axiomatize their relationship, i.e., the notion of belonging.

Wait! What?

To tell the truth, this was done long ago, in the 19th century by Georg Cantor, as we shall explain in Chapter 3, and the only difference with what we assert and what Georg Cantor asserted is that we are using the terms data and datum, whereas he used the concepts of set and element, respectively.

To put it bluntly, data is to datum as set is to element.

It naturally follows that any mathematically legitimate notion of data algebra must be couched in terms of (and subject to the rigor of) set theory and be independent of any preconceived notions about data: visual, structural or otherwise.

We hope to explain, as simply as we can, the algebraic tale of data. It is a tale of sets recounted by ascending towers of sets of ever increasing cardinality of, in the words of Paul Halmos, "frightening height and complexity."

But don't be too concerned; all algebraic tales are like that.

A final note

This book adopts the style of ending this chapter and each one that follows, with one or more pages entitled Taking Stock. They are provided so that readers can go to these pages to remind themselves of the major points covered in the chapter.

And so...

Taking Stock

The following bullet points summarize this chapter:

- This book is complemented by the Algebraix Library, a Python library containing definitions and functions that implement data algebra. The URL is http://algebraixlib.readthedocs.org/en/latest/.

- When applied, mathematics is useful by virtue of its ability to model aspects of the real world. The data algebra described in this book accurately models a computer's use of data.

- The only thing computer software ever does is take a collection of data and transform it into another collection of data.

- Algebras are useful because of their abstraction and generalization capabilities. Naturally, this is true of data algebra

- The question "What is data?" is, on reflection, not such a simple question to answer. Our initial response to this question is to note that the natural habitat of data is set theory.

- But if this is so, how do we account for the apparent fact that real-world data is neither homogeneous nor necessarily algebraic? Indeed, it seems to be an unruly mess compared to one's usual perception of a well-behaved mathematical space. We consider this question in greater depth in Chapter 5.

Chapter 2: In Search of a Data Algebra

We all want progress, but if you're on the wrong road, progress means doing an about-turn and walking back to the right road; in that case, the man who turns back soonest is the most progressive.

~ C. S. Lewis

———— ❦ ————

THE ALGEBRA OF DATA was invented by one of the authors of this book, a mathematics professor who was an early employee of Algebraix Data Corporation, a company that was established to develop software products and services based on data algebra. The founders of Algebraix Data were software engineers.

At the time the company was founded, the notion of a data algebra was only a collective gleam in the founders' eyes. The data algebra that was gradually built and mathematically tested in a variety of data processing contexts took about 5 years to evolve. In parallel with this, the company worked on building data management technology that employed the evolving algebra.

The fact that a data algebra never existed until a handful of years ago will likely surprise you. If you are not a software engineer, you probably assume that such an algebra was developed many decades ago and is regularly employed by software engineers to resolve tricky engineering problems they encounter. Alternatively, you may be a software engineer, in which case you may believe that the universe of data, in its extraordinary variety, is simply too complex and convoluted for mathematics to be usefully employed. Both these perspectives are wrong.

The original concept of what a computer could be was mathematical. It was a concept invented by mathematician, Alan Turing, in 1936. Later, he made important contributions to the design of the electromechanical and electronic code breaking machines that were built and successfully employed during the Second World War. These machines were, in reality, single purpose computers.

However, in the wake of the Second World War, mathematics played only a subsidiary role in the evolution of computers. Most of the advances in computer hardware and computer software that occurred in those years were achievements of electronic engineering. This unfolded in the following way.

In 1945, mathematician and physicist John von Neumann (and others) wrote a paper entitled *First Draft of a Report on the EDVAC*. EDVAC was an acronym for Electronic Discrete Variable Automatic Computer. During the war, Von Neumann had also been working on difficult mathematical problems, particularly the calculation of artillery tables to better enable the effectiveness of artillery. Electronic calculators had been used to great effect for that purpose.

In the paper, von Neumann described what has come to be called "The Von Neumann computer architecture." This pragmatic architecture involved a processing unit embodying an arithmetic logic unit and processor registers, a control unit containing an instruction register and program counter and memory to store both data and instructions. It also included external mass storage and input and output mechanisms for data. The model, rather than an architecture based on a Turing Machine, established itself at the foundation of the computer industry that gradually emerged.

Mathematicians had some involvement in the further evolution of computers. But there were few efforts to model the processing of data in a mathematical fashion, with data conforming to elements or collections of elements within an algebraic domain and program instructions conforming to algebraic operations on that data – and those efforts were unsuccessful.

Shaving the barber

The barber problem is also known as Russell's paradox, since the famed philosopher and mathematician, Bertrand Russell, popularized it. He used it as a means of pointing out a fundamental problem in Georg Cantor's definition of set theory. The barber problem is this:

> A regiment in the army has a designated barber. The colonel in control of the regiment gives orders to the barber. He commands thus, "From today, every man in the regiment shall be clean-shaven. Henceforth you must shave every man who does not shave himself."

The problem is: who shaves the barber?

Logically, the barber cannot shave himself since he is commanded to shave men who do not shave themselves. And yet if he does not shave himself, he becomes one of the men whom he should shave.

The barber problem is a variation of a familiar Greek paradox (the circular "All Cretans are liars. I am a Cretan"). Its relevance to set theory is that when set theory was originally created by Georg Cantor, a set was deemed to be any definable collection. Russell demonstrated that sets needed to be defined more precisely than that, by considering those sets that are "members of themselves." The problem, which is identical to the barber problem, can be described mathematically as follows:

> Let R be the set of all sets that are not members of themselves. If R is not a member of itself then by definition it must contain itself. But if it contains itself then it violates its own definition as the set of all sets that are not members of themselves. We can express this algebraically as follows:
>
> Let $R = \{A: A \notin A\}$, then $R \in R$ if, and only if, $R \notin R$,
>
> i.e., let R equal the sets A, such that A is not an element of A, then "A is an element of itself" implies "R is not an element of itself" and vice versa.

Mathematically, the consequence of the barber problem was that the axioms of set theory were reviewed and changed so that such paradoxes were eliminated. Subsequently, set theory became a foundational branch of mathematics, partly because most mathematical objects could be represented within set theory. All data, whatever form it takes, can also be represented as a set or an element of a set.

Functional programming

Despite the computer's evolutionary path, computational mathematics did not die simply because the von Neumann architecture became dominant. In fact, a good deal of work was done on what came to be called lambda calculus (λ-calculus), which in turn played an important role in the theory of programming languages.

We will not describe lambda calculus here, but nevertheless we feel obliged to draw attention to the Church-Turing Thesis, which states that "the untyped lambda calculus is capable of computing all effectively calculable functions." Put simply, it means that mathematicians have a computational model that they believe to be complete.

A natural consequence of the invention of lambda calculus was its use in development of various programming languages. Such languages are generally referred to as functional programming languages, and they resemble algebraic activity by virtue of their applying functions to carry out calculations and transformations. In other words, functional programming languages process data in a mathematical way. What they are missing is a data algebra to define the data.

The tower of Babel

Almost everyone who enters the software world quickly becomes astonished at the number of different programming languages that exist. It has even been suggested that there are now more programming languages than there are human languages. For the record, there are about 6000 human languages, but only about 100 of those languages are spoken by significant populations of people.

The same could be said of programming languages. There are certainly thousands. Wikipedia[3] lists 600 or so languages that it regards as notable. If you read through the list you may encounter many that you have never heard of before, and yet there are many many more, thousands more, that never made the cut.

Programming languages proliferate for a variety of reasons. Often innovations in computer hardware provoke the creation of new languages. Sometimes the need to specialize a language for a particular area of application (statistics, graphics, video rendering, text processing, etc.) forges a new language. Sometimes software companies wish to create a proprietary language they can control. Sometimes it's simply a matter of evolution. New problems are encountered, new features are added, and thus new dialects are born. Some languages (notably object-oriented ones) are extensible, and hence they evolve of their own accord.

The evolution of functional languages

The first language with functional features was Lisp, developed by John McCarthy (at MIT) in the late 1950s. Other languages such as IPL and APL followed in its wake. In 1977, John Backus formally introduced the idea of

---

3   See http://en.wikipedia.org/wiki/List_of_programming_languages

functional programming in a Turing Award lecture entitled: *Can Programming Be Liberated From the von Neumann Style? A Functional Style and its Algebra of Programs*. He defined functional programs as being constructed in a hierarchical way by means of "combining forms" that allow an "algebra of programs." His paper generated enough interest to provoke research into functional programming.

This gave rise to a whole raft of new programming languages, including: ML, SASL, Miranda, NPL, Hope, J, K and Q. Then, in 1987, there was a general move to form an open standard for functional programming, and the Haskell language was born. Haskell, an open source language, has been evolving ever since then.

Functional programming languages work in a particular way. They are *declarative*. This means that the program tells the computer what to do without telling it how to do it. This is in contrast to the procedural or imperative style of programming, where the language specifically controls the flow (the procedure) of the computer's activity.

Declarative is an umbrella term in the sense that it covers a number of different language behaviors. In the case of functional programming languages, the language implementation tries to ensure that there are no side effects to the execution of any functions (bugs in procedural programs often turn out to be unexpected side effects of the program code).

So a functional program is a collection of functions that will be executed as specified by the programmer. Ideally, all state changes to any collections of data will be represented as functions. All of this puts an onus on the compiler to resolve how the computer will execute the program. However, it should make it much easier for a programmer to understand and predict the behavior of a program he writes.

Hybrid languages

Although the solution of programming problems may in theory be best addressed by thinking of the problem in algebraic terms and formulating a solution by writing algebraic functions, with some applications, it is really quite difficult for programmers to write code in that manner. This has caused a natural objection to a functional programming language.

For example, algebraic functions are natural features of the R language, which focuses on statistical calculations and algorithms, and the Julia language, which focuses on scientific and high performance computing. It is less obvious why you would need such capability in languages like Perl, Javascript, Visual Basic or even the almost defunct COBOL. All are general purpose languages that are more likely to be used for text processing or transactional applications than algebraic manipulations.

Perhaps you can get the best of both worlds with a language that implements function calls after the fashion of a functional programming language but also has procedural features for handling situations where writing procedural code (telling the computer how to do something) is easier for the programmer.

There are, in fact, many such hybrid languages, including Python, Ruby, Scala, Java and Erlang. Among these, Python is the most popular, which is one of the reasons that Algebraix Data Corporation has chosen to use it as the basis of its Algebraix language.

And Yet, No Data Algebra...

Our brief foray into the evolution of functional programming languages shows that there have been pioneers who tried to exploit mathematical ideas in software. It is well known that the dominance of the von Neumann architecture led to a less mathematically oriented computer industry, but what the practical consequences were and what to do about it has never been clear.

The enthusiasm for functional programming was and is an enthusiasm for mathematics. It was fostered by individuals who were convinced that there were significant advantages to a mathematical style of programming. And there is virtue in it. When such a programming style is enshrined in a programming language, and competently employed, it gives rise to fewer programming errors, requires fewer lines of code and promotes software reuse. In short, it can improve productivity significantly.

Nevertheless, while there have been efforts to mimic mathematics in software, there has been no successful attempt to define an explicit algebra of data that enables data to be dealt with in a mathematical fashion.

The relational malaise

It can be argued that the relational database movement squashed the possibility of data algebra for many years by a process of camouflage. It was born of good intentions. As far as we can tell from the historical record, it was a genuine attempt to introduce a data algebra – the so-called "relational algebra." However, if it ever had a sound mathematical foundation, it soon sacrificed it.

The fundamental idea underlying relational database is to organize data into tables (often called relations) of rows and columns. In general, every entity (a thing: person, product, order, invoice, etc.) has its own table and the various attributes that describe it are columns in the table. Each row in the table represents one instance of the entity (one specific person or product or whatever). In practice, not all tables represent entities, but those that do assign a unique key to each row. This is a physical implementation tactic that ensures the uniqueness of rows, and is useful for linking one entity to another (for instance linking a customer to an order). When used in this way, the unique key is described as a foreign key and is often used to join tables together.

All of this works effectively for data that fits well into tables. If relational database theory had confined itself to managing data in tables, then it is possible that a valid, but very limited, algebra could have been constructed around it. It could, perhaps, have based itself on matrix algebra (a matrix is a table or rectangular array of numbers, symbols, or expressions, arranged in rows and columns).

Unfortunately, that possibility was demolished by the introduction of SQL as a standard query language and, particularly, by the introduction of the null value, a wholly unmathematical idea. This deserves a little explanation.

Practically speaking, in a relational database, a null value can mean any of the following things:

- The value does not exist. One example might be the Social Security Number (SSN) of a U.S. resident who has not yet been issued with such a number – a newborn baby, for example.

- The value was not provided. An example could be the SSN of a U.S. resident who chooses not to provide such information.

- The value is not yet known (i.e., it could be anything). An example is the SSN of a U.S. resident who has not yet provided the database with a value for this attribute.

- The attribute is not applicable to this instance. An example is the SSN of someone who is neither a U.S. citizen nor a U.S. resident.

The first three of these situations could be kluged by assigning specific values to indicate what the context of the missing value is. And it is indeed a kluge since it changes the assigned meaning of the data item SSN. It would now mean "either an indicator denoting the status of this item or an SSN." An alternative kluge would be to assign a zero value to all three, bundling all three of these meanings together under the heading of "we do not know."

The last four null possibilities are far more troubling. If an entity has attributes that do not apply, then it genuinely does not have those attributes. For example, if we create a table of "living things" and include as an attribute of this table "color of beak," then many rows in the table (almost all mammals) will have null values for this attribute.

Thus, a table is the wrong structure to represent such a collection of data. If you suspect that this is a little synthetic as an example, consider instead a product table for a building materials retailer. Create such a product table, and many products will have attributes that do not apply. Paint may be sold by specific color, an attribute that is not applicable to many other products. Wood may be sold by length, thickness and width. Sand and cement may be sold by weight, and so on. There is undoubtedly an entity called product, but it cannot be represented accurately as a table. When forced under duress into a table, the table becomes populated by a great number of nulls.

If we wanted to represent this mathematically, the nulls would have to disappear, because they represent something that doesn't exist – and with their disappearance, the convenient table structure would also disappear. Instead it would become a collection of products with varying attributes.

The null is schizophrenic, possibly indicating an absent value or possibly indicating that the attribute simply does not apply. This fundamental mathematical error might not have had such a damaging impact if it were not for SQL. SQL had to deal with the nulls. In doing so, it invented a variety of rules which forced relational database further and further away from mathematical legitimacy.

These were not the only failings of the relational approach. Tables are far too limiting to be the only allowable data structures. Thus, other databases emerged (document database, XML database, RDF database, etc.) to cater to other common data structures (hierarchical data, text, graphs, etc.). However, these databases have no mathematical legitimacy either.

The consequences of bad algebra

We could make many criticisms of the relational database from a mathematical standpoint, but we think we have said enough. In the early days of the relational database movement, advocates for the technology were even emboldened to proclaim that the relational approach to data was mathematically correct, when in reality, it wasn't mathematical at all.

When data could be shoe-horned into tables, relational databases worked reasonably well. When the data didn't fit so nicely, there were problems. The idea that there is a right way to structure data (i.e., in tables) is preposterous. In truth, there are simply useful ways to structure data. Sometimes data is best structured as a graph (such as a network of relationships), sometimes as a recursive structure (such as a bill of materials), sometimes as an ordered list, and so on. Relational databases do some of these things very poorly.

Soon after relational databases became dominant, object-oriented programming languages came to the fore. These languages, which soon became quite popular, did not represent data in a relational fashion at all. This created an awkward problem when programmers chose to store data. They were obliged to translate the data as they had represented it in their program into the form the database required. This problem was called the impedance mismatch. It was so prevalent that an open source mapping capability, Hibernate, was written and widely applied to address the problem.

Thus, an IT industry emerged where programming languages and databases were incompatible in the way they modeled data. Clearly such a situation could not persist indefinitely. Nevertheless, the initial attempt to establish object databases (to complement object-oriented languages) failed commercially, and it wasn't until a second generation of object databases – this time called document databases – emerged that relational databases experienced any competition. However, these relatively new databases have no foundation in a data algebra either.

Axiomatic extended set theory

When E. F. Codd set out to formulate relational theory, he based some of his ideas on extended set theory, which was an idea formulated and described in a 1968 paper, *Description of a Set-Theoretic Data Structure* by D. L. Childs. Childs correctly identified set theory as the natural mathematical basis for representing data and proposed an extended set theory as a means to creating an algebra of data. As a consequence, it would be possible to query data using set operations such as union, intersection, Cartesian product and so on. If this were done, the use of sets and set operations would provide complete independence from physical data structures.

When Algebraix Data Corporation began work on building a data algebra, axiomatic extended set theory, as expounded by D. L. Childs, was examined to see if it was fit for this purpose. It was soon concluded that it constituted a misguided effort.

The assumption on which the genesis of data algebra moved forward was that it should be possible to derive data algebra directly from Zermelo-Fraenkel set theory (ZFC) without need for any extension. This proved to be correct. The algebra of data described and explained in this book does indeed derive directly from ZFC. Childs' extension was unnecessary.

Nevertheless, teasing out such a universal data algebra from ZFC was not plain sailing. If it had been, an algebra of data would have been propounded decades ago, and it would quickly have become one of the foundations of software.

A final thought

Our intention effort in this chapter was to provide a brief review of where mathematics has been employed in software engineering at a foundational level. There are, of course, many programs and software products that implement mathematical techniques and formulae to process data (mainly numerical data) in many ways to excellent effect. Functional programming languages were developed to implement mathematical functions in a formal manner. Yet, despite the fact that data is at times even discussed in terms of sets, there was no successful attempt to implement a set theoretical approach to data – until recently.

Taking Stock

The following bullet points summarize this chapter:

- While born of the mathematical work of Alan Turing, the computer industry did not evolve mathematically. Following the computer architecture proposed by John von Neumann, it soon became the domain of electronic engineers and software engineers.

- There were some attempts to introduce and leverage mathematical ideas in software. They were as follows:

  - The introduction of functional programming languages, the most popular of which is Haskell. Functional programming languages implement a programming style that treats computation as the evaluation of mathematical functions while avoiding changing-state and mutable data.

  - E. F. Codd's so-called relational algebra.

  - D. L. Child's extended set theory.

- None of this resulted in the formulation of an algebra of data. Not even close.

Chapter 3: Cantor, Zermelo and Fraenkel

...we shall study sets, and sets of sets, and similar towers of sometimes frightening height and complexity, and nothing else.

~ Paul R. Halmos

⸻

GEORG CANTOR, a German mathematician, invented set theory single-handedly. In 1874, he unleashed it on the world by way of an article in **Crelle's Journal**. It sparked immediate controversy but very quickly had an impact, and it was soon accepted. It became very clear that set theory was fundamental to the whole discipline of mathematics. As a direct consequence, mathematicians around the world immediately set to work trying to retrofit set theory into other branches of mathematics.

As we noted in the previous chapter, Cantor's formulation of set theory was not perfect, or better put, it was not free of paradoxes. It needed to be massaged a little for the sake of rigor, so in the early twentieth century, various axiomatic systems were proposed to fix the situation. The system suggested by Ernst Zermelo and Abraham Fraenkel was eventually adopted, and for that reason, set theory is often referred to as Zermolo-Fraenkel set theory, which may seem a little unfair given that it was Cantor's invention.

The algebra of data derives from set theory. To be precise:

The algebra of data is applied set theory!

For that reason, we have chosen here to describe and explain all the axioms of ZFC set theory on which data algebra is based. We have two motives for this.

1. We believe this will give you an opportunity to become familiar with the mathematical symbols and terminology used in set theory before we dive into the algebra of data for real.

2. These axioms are important in the derivation of data algebra, and thus it will serve you to become familiar with them. In particular, it will help if you understand what a power set is.

Given that we are about to introduce you to a whole series of mathematical symbols and notation, now is probably a good time to provide a complete

21

list of every symbol used in the book in various data algebra expressions. You may be familiar with some of these but will not be familiar with all of them. We will explain the unfamiliar ones as they arise.

Mathematical symbols	
Symbol	Description
{ }	Braces indicate sets
:=	Is defined as
∈	Belongs to
∉	Does not belong to
⊂	Subset of
⊊	Subset of and not equal to
| |	Cardinality
∘	Composition
↔	Transpose
X	Cartesian product
∅	The empty set
∪	Union
∩	Intersection
′	Complement
−	Difference
▼	Cross-union
▲	Cross-intersection
▷	Superstriction
◁	Substriction
►	Cross-superstriction
◄	Cross-substriction
⋈	Natural join

Here is also an appropriate time to draw your attention to the notations we use to distinguish between various mathematical objects. We also provide a list of these in tabular form, along with font descriptions.

Notation standards		
Item	*Convention*	*Example*
element of a set	regular italic	a
couplet	regular italic with superscript	a^b
set	regular italic	A
set of sets	bold: font = Gauss	$\mathcal{A}$
genesis set	bold: font = Gauss	$\mathcal{G}$
power set	bold: font = Fraktur	$\mathfrak{P}$
sets of numbers	regular: font = Fermat	$\mathbb{N}$
relation	bold italic	$\boldsymbol{R}$
clan	bold: font = Fermat	$\mathbb{C}$
function	bold italic	$\boldsymbol{f}$
partition	regular italic: font = Greek	$\mathit{\Pi}$

A parade of axioms

Any game evolves subject to rules which are enforced by referees who, quite naturally, use language and terminology that comes from the rules. A casual game may have implicit rules, and it is usually refereed informally by the participants in a haphazard manner using idiomatic language. Examples of such games include hopscotch, politics and, in the authors' opinion, so-called relational algebra.

However, set theory is not a casual game: the rules – the axioms – are explicit, and the game is refereed by first order logic.

The primitive (undefined) notion of set theory is that of belonging. If an element e (whatever e is) belongs to a set E (whatever E is) we write:

$$e \in E.$$

We may also describe this notation in words by saying that "*e is an element of the set* E" or "*e is contained in* E." To assert that "*e is not an element of the set* E" we write:

$$e \notin E.$$

23

This may seem all hunky dory, but the very mention of E begs questions, such as "*How do we know that such a thing as a set exists,*" and "*If it does, what is it?*" Enter the axiom of existence.

Axiom of existence: *There exists a set.*

So, maybe it's E, maybe it isn't.

Axiom of extension: *Sets A and B are equal, denoted by A = B, if, and only if, they have the same elements.*

This axiom suggests the possibility of other relationships between sets aside from equality. These are the possibilities:

- If A and B are not equal, we write $A \neq B$.

- If $a \in A$ implies $a \in B$; then we say A is a subset of B (or B is a superset of A) and we write $A \subset B$. It follows that $A = B$, if, and only if, $A \subset B$ and $B \subset A$.

- If $A \subset B$ and $A \neq B$, then we write $A \subsetneq B$ and refer to A as a **proper subset** of B.

We illustrate these relationships in Figure 1, where the circles represent sets and the lowercase letters represent elements.

If our previously mentioned set E consists of exactly three distinct elements, say a, b and c, we use the visual artifice $\{a, b, c\}$ to represent E and we write $E = \{a, b, c\}$ (note the subtle point here that any kind of representation, such as a visual representation – which is exactly what we mean by visual artifice – is **not** the mathematical object, just a way of looking at it). The $\{a, b, c\}$ is the standard notation for a set. The elements are given, separated by commas, and the braces mark out the beginning and the end of the parade of elements.

We can deduce from the axiom of extension that $\{a, c, b\}$ and $\{c, b, c, a, c, a\}$ are also representations of exactly the same set E. This might not be immediately obvious, but the axiom of extension confers equality irrespective of the order of the elements. So $\{a, c, b\} = \{a, b, c\} = E$. Also, if any element is repeated, it still only counts once – further mentions of the element are redundant. So $\{c, b, c, a, c, a\} = \{c, b, a\} = E$. The cardinality of E (the number of elements in E) is three, and we write $|E| = 3$.

So how do we construct the subsets of a set? What are the rules?

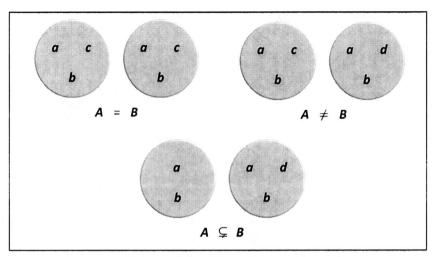

Figure 1: The axiom of extension

Axiom of specification: *Every set A and every statement S(x) determines a unique subset S of A whose elements are exactly those elements x of A for which S(x) holds.*

To indicate mathematically how S is generated from A by $S(x)$ we write:

$$S = \{x \in A: S(x)\}.$$

It turns out that there are a lot of ways of mathematically specifying a statement $S(x)$. You can use atomic statements of belonging, such as $x \in E$. You can use statements of equality, such as $A = B$, and you can combine collections of such statements using the syntax of all the logical operators. The full inventory of logical operators is: $\vee$, $\wedge$, $\neg$, $\Leftrightarrow$, $\Rightarrow$, $\forall$ and $\exists$. However, we have chosen not to employ any of these in this book, preferring to use the equivalent words instead. If you do encounter any mathematical symbols you are unfamiliar with, there is a reference in Appendix A, which lists them all with descriptions. It includes every new, familiar or arcane mathematical term we employ in this book.

Consider the following example, where we use the set B to generate a set S from the set A:

$$S = A - B = \{x \in A: x \notin B\}.$$

This set S is called the difference between A and B. We have introduced "$-$" the difference operator, so S is the difference of the two sets, which means the set of elements that belong to A that do not belong to B.

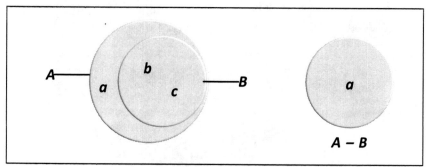

Figure 2: The difference of A and B, A − B

A surprisingly useful, if somewhat slippery, set emerges as a consequence of these three axioms. Consider the set $\{x \in E: x \neq x\}$. Of course, there is no element in the whole set-theoretical universe that is not equal to itself. Nevertheless, this is not mathematical nonsense; it is an effective way of defining a set with nothing in it.

It follows from the axiom of extension that this set, which we denote by ø and refer to as the **empty set**, is in fact unique. Moreover, ø has the surprisingly useful property of being a subset of any set.

But hold on a moment, can a set be an element of a set? Of course it can, but we need to have a rule to state that it is so.

Axiom of pairing: *For sets A and B the set {A, B} exists.*

This axiom allows us to play with the empty set ø in a truly productive way.

Consider the situation where $A = B = ø$. We can immediately, and dramatically, enlarge our inventory of sets to include:

$$\{ø\}, \{ø,\{ø\}\}, \{ø,\{ø,\{ø\}\}\}, \{ø,\{ø,\{ø,\{ø\}\}\}\}, ...$$

And this sequence of sets hints at the genesis of the set of natural numbers:

$$\mathbb{N} := \{1, 2, 3, ...\},$$

as long as we take the ellipsis, (the ... in each set) to mean "ad infinitum."

In fact, there is a mathematical method for generating the natural numbers $\mathbb{N}$ from the empty set, but we will not burden you with that. However, in case you never noticed, we slipped infinity into the mix here, so we are obliged to note that there is an **axiom of infinity,** which states "*there is an infinite set.*"

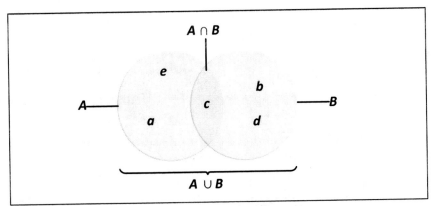

Figure 3: Union and intersection

This leads us on to the **axiom of choice,** which, in simple words, states: *"one may construct a set from an infinite collection of sets by choosing an element from each set of the collection."*

This then leads us to the **axiom of substitution,** which roughly states: *"anything intelligent that one can do to the elements of a set yields a set."*

If you are starting to get the impression that we are dashing through a bad part of town hoping to get through it without mishap, then you are exactly correct. On the back of such axioms, we could digress into long, set-theoretic philosophical discussions of the implications of infinity and how it should or should not be handled. We succeed in evading that discussion with the convenient excuse that: "computers are finite, and therefore for all practical (computer) purposes, the algebra of data only needs to deal with finite sets."

So, subject to the caveat that, from time to time, it is clerically convenient to view a particular finite set as a subset of some countably infinite set – a set whose cardinality is the same as the cardinality of $\mathbb{N}$ – we get to avoid the subtleties of infinite sets.

Let's move on to another fundamental question: "Can a set be a subset of a set other than itself?" Of course it can.

Axiom of union: *For sets A and B, there exists a set U such that $A \subset U$ and $B \subset U$.*

Incidentally, we generally use U to denote "the set of elements under consideration." So for example, if we are considering data about people, then we would most likely use U to denote a large population of people in a state or even in a country. It's merely a convention, of course.

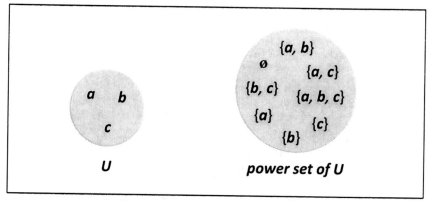

Figure 4: U and the power set of U

The axiom of union enables us to define the union operator, $\cup$, by way of the axiom of extension. The axiom of extension implies that the set

$$\{\, x \in U : x \in A \text{ or } x \in B \,\}$$

is unique. We refer to this set as the **union** of A and B, and we denote it by $A \cup B$. With the union of A and B in hand, we can create the subset

$$\{\, x \in A \cup B : x \in A \text{ and } x \in B \,\}.$$

This set is referred to as the **intersection** of A and B, and it is denoted by $A \cap B$. And, if $A \cap B = \emptyset$, we say A and B are **disjoint**.

So now we have acquired two very useful binary operators for manipulating sets. The next question we can ask is:

What is the set-theoretic universe of sets that these binary operators can act within? What is the universe within which they live?

If you pursued an education that focused on science or engineering, and included a healthy dose of mathematics, it is quite possible that every set-theoretic idea we have introduced so far is not new to you. However, what we are about to describe may be new, and even if it isn't, how we are going to use this last axiom, probably is. So please take note.

Axiom of power: *For each set U, the set consisting of the subsets of U exists.*

The set of all the subsets of U is denoted by $\mathfrak{P}(U)$ and is called the power set of U. Naturally, the root set or genesis set is denoted by U to suggest "universe" because it contains all of the elements necessary to create $\mathfrak{P}(U)$.

We need to emphasize here that U is very definitely a local universe, because there is no set that contains everything. If such a set existed, it would invite the wrath of the Russell paradox (see *Shaving the barber* in Chapter 2) as it would contain an element that it does not contain!

A final note

The word "set" is a very overloaded word. It's the English word that has more definitions than any other – 464 definitions in all, according to *The Oxford English Dictionary* (if you count all variations of nouns, verbs and adjectives). One of those definitions attempts to explain the mathematical sense of the word. However, it is so misleading that we have chosen not to include it here at all. As one might suspect, other dictionaries are no better.

We tentatively suggest that if dictionary compilers feel obligated to dip their toe in the mathematical water, they adopt a definition along the lines of: *a set is a mathematical structure representing a collection of things that is rigorously defined by the ZFC set theory axioms.*

The Power of Ascension

You may not have noticed that, so far, we haven't said much about what an algebra actually is in set-theoretic terms. Realistically, we had to cover several miles of axiomatic territory before we could do that. But now that we are armed to the teeth with the concept of the power set, we can advance in that direction.

The word "power" in the term "power set" derives from the fact that if U has exactly n elements, then $\mathfrak{P}(U)$ has exactly 2^n (two to the power n) elements. And, as we can also create $\mathfrak{P}^2(U) := \mathfrak{P}(\mathfrak{P}(U))$, if we compute $2^{(2^n)}$ we will know exactly how many elements there are in the power set of the power set. And we can keep applying power set to power set as many times as we care to.

Let's take a quick look at a simple power set. If $U = \{a, b, c\}$ then:

$$\mathfrak{P}(U) = \{\emptyset, \{a\}, \{b\}, \{c\}, \{a, b\}, \{a, c\}, \{b, c\}, \{a, b, c\}\}.$$

U had 3 elements and $\mathfrak{P}(U)$ has 8 (i.e., 2^3) elements. A clear advantage of the move from U to $\mathfrak{P}(U)$ is the exponential increase in the number of toys in the toy box. Who could object to more toys, especially when the increase is from n to 2^n?

Well, to be perfectly honest, you might not be entirely happy with this when you discover you have lost some of your old favorite toys. You see:

$$\text{for every } a \in U, a \notin \mathfrak{P}(U).$$

To put it bluntly, you appear to have lost all your old toys, because none of them can be found in $\mathfrak{P}(U)$. But don't be too concerned; like all the Toy Story movies, this is a toy story with a happy ending.

Before we reveal that, it is probably best that we define the term "ascension." It is an important term and an important idea. We never mentioned it before because we needed you to lose all your toys in order to have a reason to explain it. If we move from operating on the set U to operating on the set $\mathfrak{P}(U)$, we say that we have ascended (or lifted) from U to $\mathfrak{P}(U)$. And that could be a truly positive thing if this newly minted power set provides us with extra power.

So first of all, we note that the toys we lost from U have been reincarnated in $\mathfrak{P}(U)$. In other words, for every $a \in U$, $\{a\} \in \mathfrak{P}(U)$. The reality is that a is not "lost" in the ascent, it is just "renamed."

Let's not forget that every element of $\mathfrak{P}(U)$ is a subset of U. This turns out to be something we can leverage. Of itself, $\mathfrak{P}(U)$ sprouts a natural unary operation which we refer to as **complementation** and define in the following way:

$$A' := U - A \text{ for each } A \in \mathfrak{P}(U).$$

It naturally follows that $A' \cup A = U$ and $A' \cap A = \emptyset$.

But, the real power of a power set stems from the fact that it provides a richer algebraic playground than U, – regardless of the nature of the elements of U – homogeneous or heterogeneous, algebraically fecund or algebraically barren. Believe it or not, by lifting from U to $\mathfrak{P}(U)$ we have gotten ourselves a bona fide algebra.

It is, in fact, a set algebra with a signature:

$$[\mathfrak{P}(U), \{[\cup, \emptyset], [\cap, U]\}, \{'\}, \{\subset\}].$$

Signatures

The purpose of a signature is to show the fundamental facts about an algebra. It's a simple idea that looks more complex than it is – what with all those brackets and braces.

Before we describe it, we need to explain the idea of an operator's **identity**. In simple arithmetic, the identity of the operator + is 0. Take any number, say x, and add 0 to it and you get the same number. So the number 0 is said to be the identity for the operator +. Similarly, in arithmetic, the identity for the multiplication operator is 1. One of the details that an algebra's signature provides is the operators that are valid for the algebra, along with the identity (if there is one) for each operator. In general, binary operators (i.e., those which operate between two elements of a set) may have identities, but unary operators do not.

In the above signature, $[\cup, \emptyset]$ indicates that, for this algebra, the union operator is valid and its identity is the empty set $\emptyset$. So, for every set A, where $A \in \mathfrak{P}(U)$, $A \cup \emptyset = A$. Similarly, for every set A, where $A \in \mathfrak{P}(U)$, $A \cap U = A$.

Given that morsel of information, the algebraic signature shown above, [𝕭(U), {[∪ , Ø], [∩,U]},{ ⊂ }, { '}], can be described as follows:

The algebra's operators shown in the signature include ∪ (union), ∩ (intersection) and '(complementation). Two of these operators are binary and have identities. The third operator, ', is a unary operator and hence has no identity (binary operators have identities, and unary operators do not and cannot).

Finally, the signature shows the relation ⊂, which indicates the subset relationship.

Note that the signature's purpose is to highlight operations and relations that are relevant to the algebraic quest at hand. Clearly we could have mentioned the "=" relationship, but it is assumed to be part of the algebra, as it is part of every algebra. Also the signature does not inform us of all the properties of the operators that it highlights.

Nor is it necessarily the last word. As we investigate a particular algebra, we may invent a completely new operator or relation to add to the signature. As such, the signature is a living entity that may well evolve in time.

The properties of the binary operators

We now need to examine the behavior of these operators within this algebraic universe. We can start with the idea of **commutativity**.

We know that in arithmetic, $3 + 4 = 4 + 3$, or generally, $a + b = b + a$. This is commutativity, and we say that addition is commutative. ∪ and ∩ are clearly commutative:

$$A \cup B = B \cup A \text{ and } A \cap B = B \cap A.$$

Union and intersection are **associative,** just as addition is:

$$a + (b + c) = (a + b) + c.$$

Thus:

$$(A \cup B) \cup C = A \cup (B \cup C)$$

$$\text{and } (A \cap B) \cap C = A \cap (B \cap C).$$

And they have the property of **mutual distributivity**. With the operators ∪ and ∩:

$$(A \cup B) \cap C = (A \cap C) \cup (B \cap C)$$

$$\text{and } (A \cap B) \cup C = (A \cup C) \cap (B \cup C).$$

If you are getting the idea that the operators $\cup$ and $\cap$ are a little like the arithmetic operators + and −, that's fine, but please do not take the comparison too far, because, aside from other differences, $\cup$ and $\cap$ are also **idempotent** but + and − are not (and also, to tell the honest mathematical truth, and this may come as a surprise to you, − is really a unary operator[4]).

"Idempotent," incidentally is a word you will not find a great deal of use for outside mathematics. Etymologically, it comes from Latin and means "having the same power" (in Latin *idem* = the same, *potentem* = powerful). Practically, what idempotence means for $\cup$ and $\cap$ is that:

$$A \cup A = A \text{ and } A \cap A = A.$$

Of flarns, clarps and tworbles

The interaction of union, intersection and complementation is neatly captured by so-called De Morgan laws:

$$(A \cup B)' = A' \cap B' \text{ and } (A \cap B)' = A' \cup B'.$$

Stated in words, this says that:

>the complement of a union is the intersection of the complements

and:

>the complement of an intersection is the union of the complements.

Here we have the opportunity to introduce the nonsense words "flarn," "clarp and "tworble." Who could resist? These words were borrowed quite a while ago from Johnny Carson's TV show by the professorial member of this book's authoring team. They are intended to have no specific meaning, aside from the fact that a flarn is not a clarp and that neither of the two are tworbles. They provide us with a useful mnemonic generalization.

Consider again: *the complement of a union is the intersection of the complements* and *the complement of an intersection is the union of the complements*. This sounds like the also-true assertion from school mathematics that: *the log of a product is the sum of the logs.*

---

4   Mathematically, the expression $a - b$ is really a shorthand for $a + (- b)$. The unary operator "−" is applied to b, then the binary operator + is applied to both terms.

Aren't all three of the previous italicized mantras specific versions of the more general: *the-flarn-of-the-clarp-is-the-tworble-of-the-flarns*?

Yes they are. But why are we asking?

Ascension redux

Let us revisit ascension. We need to do this because we will be using ascension several times as we gradually walk you through the levels of the tower of data algebra.

So consider our set U, the genesis set of elements we start out with. We can infect U with some algebra by taking it to be your very favorite number system from elementary school: $\mathbb{Z}$, the set of integers. The integer algebra you are familiar with is summarized by the signature

$$[\mathbb{Z}, \{[+, 0]; [x, 1]\}; \{-\}, \{<\}].$$

In words, this signature says that the set of integers has two binary operators $+$ and x and a unary operator $-$ and a relationship $<$. We can ascend from this algebra by taking the power set of $\mathbb{Z}$. And, as we mentioned but did not prove a few pages ago, ascending from U to $\mathfrak{P}(U)$ yields a Boolean algebra with signature:

$$[\mathfrak{P}(\mathbb{Z}), \{[\cup, \emptyset], [\cap, U]\}, \{'\}, \{\subset\}].$$

In view of what we already discussed, we can legitimately ask, "Have we gained set algebra at the expense of integer algebra?"

Well, yes, we have. The integers have vanished. So, $2 \notin \mathfrak{P}(\mathbb{Z})$, $3 \notin \mathfrak{P}(\mathbb{Z})$ and so on. Nevertheless, they are included in a way, because $\{2\} \in \mathfrak{P}(\mathbb{Z})$, $\{3\} \in \mathfrak{P}(\mathbb{Z})$. What is more, $\{2 + 3\} = \{5\}$ and $\{2 \times 3\} = \{6\}$. So we can chose to designate the $+$ and x operators to work in the following way:

$$\{2\} + \{3\} = \{5\} \text{ and } \{2\} \times \{3\} = \{6\}.$$

Be clear about what we have done here. The operators $+$ and x in the integer algebra we started with are not the same as the operators $+$ and x in the Boolean algebra of $\mathfrak{P}(\mathbb{Z})$. We have "overloaded" these two symbols. But it doesn't cause any problems, because $\{\{i\} \in \mathfrak{P}(\mathbb{Z}): i \in \mathbb{Z}\}$ perfectly replicates the algebra of $\mathbb{Z}$ in $\mathfrak{P}(\mathbb{Z})$.

It may be clearer and perhaps even instructive to translate this into words: *the overloaded sum (product) of two singleton sets of integers is the singleton set of the integer sum (product) of the integers.*

This generalizes in a convenient way to the elements of $\mathfrak{P}(\mathbb{Z})$, which are not necessarily singletons sets: for $A, B \in \mathfrak{P}(\mathbb{Z})$

$$A + B := \{a + b \in \mathbb{Z} : a \in A \text{ and } b \in B\}.$$

So: *the sum of sets is the set of sums* (a version of our mantra, *the-flarn-of-the-clarp-is-the-clarp-of-the-flarns*), and

$$A \times B := \{a \times b \in \mathbb{Z} : a \in A \text{ and } b \in B\}.$$

And: *the product of sets is the set of products* (again, *the-flarn-of-the-clarp-is-the-clarp-of-the-flarns*).

Let's take a look at some examples of this:

$$\{1, 3\} + \{4, -1, 0\} = \{1 + 4, 1 + (-1), 1 + 0, 3 + 4, 3 + (-1), 3 + 0\}$$
$$= \{5, 0, 1, 7, 2, 3\}$$

and:

$$\{1, 3\} \times \{4, -1, 0\} = \{1 \times 4, 1 \times (-1), 1 \times 0, 3 \times 4, 3 \times (-1), 3 \times 0\}$$
$$= \{4, -1, 0, 12, -3, 0\}.$$

It follows that $\mathfrak{P}(\mathbb{Z})$ is an "arithmetic" algebra in its own right, constructed by lifting the operations of $\mathbb{Z}$ to $\mathfrak{P}(\mathbb{Z})$, with signature

$$[\mathfrak{P}(\mathbb{Z}), \{[+, \{0\}], [\times, \{1\}]\}, \{-\}].$$

Right now it may look to you as though we've pulled off a dramatic ascension and lifted the whole of arithmetic algebra into $\mathfrak{P}(\mathbb{Z})$. But actually, we have not.

We have lifted much of the structure of integer arithmetic but not all of it. On the positive side, for $A, B, C \in \mathfrak{P}(\mathbb{Z})$ the lifted versions of + and × are commutative, associative and × distributes over +. Of course, the lift of + does not distribute over × in $\mathfrak{P}(\mathbb{Z})$, but it didn't do that in $\mathbb{Z}$ either. Indeed, it looks as though we've tamed the whole beast; that is, until the beast suddenly goes feral.

Consider: $-\{1, 2\} + \{1, 2\}$. Surely that's equal to $\{0\}$. Well, not exactly...

$$-\{1, 2\} + \{1, 2\} = \{-1, -2\} + \{1, 2\}$$
$$= \{-1+1, -1+2, -2+1, -2+2\}$$
$$= \{0, 1, -1, 0\}$$
$$= \{0, 1, -1\}.$$

Quite clearly:

$$-\{1, 2\} + \{1, 2\} \neq \{0\}.$$

This means a whole cancellation law goes missing when you lift to $\mathfrak{B}(\mathbb{Z})$. In other words, if $-A + A \neq \{0\}$, then $A + B = A + C$ does not imply that $B = C$. You cannot cancel the A from each side of the equation.

Nevertheless, the benefits of the lifting process far outweigh such anomalies. In this instance, if we look hard enough, we will discover the complete inventory of modular arithmetics in $\mathfrak{B}(\mathbb{Z})$!

The modular arithmetic of the power set $\mathfrak{B}(\mathbb{Z})$

You learned modular arithmetic, at least in a primitive way, when you were a child and learned how to read the time from a clock face. If someone asked you at 11 o'clock what time it would be in three hours, you swiftly answered "2 o'clock," adding 3 to 11 and subtracting 12. You performed modulo 12 addition.

So let's consider the simplest modular arithmetic of all: arithmetic modulo 2. And let's denote all the even integers by E and all the odd integers by O. We can confidently state that $E, O \in \mathfrak{B}(\mathbb{Z})$.

Why? Well, $\mathfrak{B}(\mathbb{Z})$ is a power set that contains all possible sets elements of $\mathbb{Z}$. Both the sets E and O must be in there somewhere.

Moreover, $E \cap O = \emptyset$ and $E \cup O = \mathbb{Z}$. So E and O partition the set $\mathbb{Z}$. Now let's focus on the subset $\{E, O\}$ of $\mathfrak{B}(\mathbb{Z})$ and start employing our operators $+$ and $\times$. Add an even number to an even number, you get an even number. Add odd to odd and you get even, and so on. The full set of possibilities for $+$ and $\times$ is shown in the two tables.

+	E	O
E	E	O
O	O	E

×	E	O
E	E	E
O	E	O

So take a look at these two tables and then replace *E* with **0**, which suggests that the common remainder when any even integer is divided by two is 0, and replace the *O* with **1**, suggesting that the common remainder of any odd integer divided by two is the integer 1.

+	0	1
0	0	1
1	1	0

×	0	1
0	0	0
1	0	1

This reveals integer arithmetic modulo 2 hiding shamefully in $\mathfrak{P}(\mathbb{Z})$, with a signature that looks like this:

$$[\{0, 1\}, \{[+, 0], [×, 1]\}]$$

In fact, the integer arithmetic modulo any-other-positive-integer-of-your-choice, including clock arithmetic, is hiding in $\mathfrak{P}(\mathbb{Z})$ somewhere.

The importance of partitions

Each of these modular algebras depends on a particular partition of Z. This is not an anomaly; partitions play a singular and prominent role in mathematics. They have a role to play in the conventional algebra you met at school and grew to love or hate, and they play a role in more general and more abstract algebras. And as you must realize, otherwise why would we even mention them, they play a role in data algebras.

So it's convenient for all of us that a partition is not a complex thing. Still, this is a game with rules, so a precise definition is in order:

> A **partition of a set** U is a set, say, *Π*, of non empty subsets of U such that the intersection of distinct elements of *Π* is empty and the union of all of the elements of *Π* is U.

And by the way, an element of a partition is referred to as a **component** of the partition.

For example, if *U* is {1, 2, 3, 4, 5}, then {{1, 2, 3}, {3, 4, 5}} is not a partition of *U* but {{1, 2}, {3, 4, 5}} is, with components {1, 2} and {3, 4, 5}. Notice also that a partition of *U* is a subset of $\mathfrak{P}(U)$, and, by the way, that means it must be an element of $\mathfrak{P}^2(U)$.

Taking Stock

In this chapter we covered a good deal of mathematical territory – all of it foundational to data algebras. The following bullet points summarize our path through that territory.

- The algebras of data are applied ZFC set theory (incidentally, **the ZFC** stands for **Zermolo-Fraenkel Set Theory** with **the axiom of Choice**).

- ZFC set theory is established by nine largely self-evident axioms. The axioms of existence, extension, specification, pairing, infinity, choice, substitution, union and power. The axiom of power – that we can form a power set consisting of all possible subsets of a local set universe – is of specific importance to data algebra.

- In set theory, the fundamental operators are union ($A \cup B$), and intersection ($A \cap B$), the unary operator of complementation $'$, where $A' = U - A$. There is also a fundamental subset relation $\subseteq$.

- Binary operators may have specific properties of commutativity, associativity, distributivity (between two binary operators) and idempotence. $\cup$ and $\cap$ are commutative, associative, mutually distributive and idempotent.

- By taking the power set $\mathfrak{P}(U)$ of a universe, U, we can construct a Boolean algebra. Algebras can be represented by signatures, and the signature of this specific Boolean algebra is:

$$[\mathfrak{P}(U), \{[\cup, \varnothing], [\cap, U]\}, \{'\}, \{\subseteq\}].$$

  Thus, we can lift U to $\mathfrak{P}(U)$.

- In lifting from an algebra to its power set we can always lift the operations of the algebra to the power set, but the algebraic properties of the lifted operations may not necessarily follow in kind. This is the case in lifting the unary operator "$-$" from $\mathbb{Z}$ to $\mathfrak{P}(\mathbb{Z})$.

- Nevertheless, in lifting $\mathbb{Z}$ to $\mathfrak{P}(\mathbb{Z})$ we discover that the whole of modular arithmetic resides within $\mathfrak{P}(\mathbb{Z})$ with each specific modular arithmetic depending on a particular partition of $\mathbb{Z}$.

- A **partition** of a set U is a set, say, Π, of non empty subsets of U such that the intersection of distinct elements of Π is empty and the union of all of the elements of Π is U.

- An element of a partition is referred to as a **component** of the partition.

- A partition of U is a subset of $\mathfrak{P}(U)$, which means it must also be an element of $\mathfrak{P}^2(U)$.

Chapter 4: Descartes' Legacy

Obvious is the most dangerous word in mathematics.

~ E. T. Bell

———ꝏ———

IN ACCORD with the academic fashion of the day, René Descartes' name was latinized to Renatus Cartesius. That, believe it or not, is why the adjective from his surname is "Cartesian" and not "Descartesian." No doubt you encountered this adjective in school mathematics, as geometry is usually introduced in terms of algebraic functions using so-called Cartesian coordinates.

This chapter commences with that, but what it really focuses on is Cartesian products, which you may not be so familiar with. However, we can commence by stumbling into some graphs that ought to inhabit your mathematical comfort zone.

"Ordered pairs"

Take a look at Figure 5. Perhaps you think: "OK, it's got a y-axis and an x-axis, and it's got some scattered Cartesian coordinates marking out some points on the two-dimensional surface referred to grandiosely, by mathematics geeks, as the Euclidian plane. And that is just a mathematics geek's playground for drawing graphs relating y to x in some obscure way and marking the resulting curve out with 'ordered pairs' plucked from $\mathbb{R}$, the set of real numbers."

Maybe you even recall the school lesson where such a graph over the Euclidian plane was first unveiled before your eyes, most probably in an effort to illustrate simple equations such as $y=2x$ and $y=x^2$. Perhaps you even recall the "vertical line test" from your school days, which is illustrated in Figure 6. The test is this: *if each vertical line that can be drawn in*

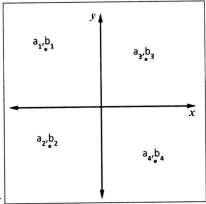

Figure 5: Cartesian coordinates

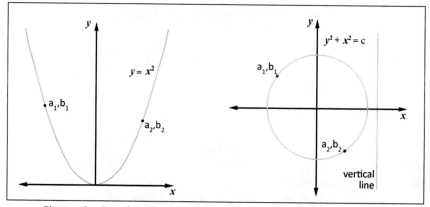

Figure 6: *Graphical representations of y = x2 and y2 + x2 = c.*

the plane intersects a curve in at most one point (an "ordered pair") the curve represents a relation between **y** *and* **x***, which is referred to as a function. If that is not so then, as in the diagram above, you have drawn a relation that is not a function.*

In some sense, functions were the gold standard of your secondary mathematics courses, running like a glittering thread through much of what you learned. However, this is not so when it comes to the algebra of data. In fact, hopefully, you will gradually realize that **relations** rather than functions are the gold standard of data algebra. Relations are gloriously flexible, and their flexibility steps forward to address data algebra's need to handle ambiguity and generality.

Now is as good a time as any to get a dark and ugly secret out into the open. If you don't already know it, then all we can say is: we hope you're ready for it...

> *Most of the relations one encounters in applying mathematics outside of the classroom do not lend themselves to such pretty closed-form expressions as $y = x^2$ and $y^2 + x^2 = c$.*

This is true in spades for data algebra. For most of the time, we will have to be satisfied with exhibiting relations as a set (of very large cardinality in practice) of "ordered pairs." You see, our data will consist of entities which aren't typically thought of as mathematical objects, even though they can indeed be represented as such objects.

By the way, have you asked yourself yet why we persist in writing "ordered pair" in quotes? If you have, good. If you haven't, do it now, and if you are up for it, try to guess, before you start reading about Kuratowski's definition of the "ordered pair." But for a few pages before that, we need to explain Cartesian products.

Cartesian products

The Cartesian product of two sets, say A and B, is a set-theoretic construction that depends on our ability to make the ascent from $A \cup B$ to $\mathfrak{P}^2(A \cup B)$, and then carefully, very carefully, as if performing brain surgery, selecting a certain subset of $\mathfrak{P}^2(A \cup B)$, which we denote by $A \times B$ and refer to as the Cartesian product of A and B. Perhaps you never quite understood that (and if you did, well done!), but don't worry, it will gradually become clear. For now, be content with the important knowledge that Cartesian products spawn the relations that make algebras tick.

Let A and B be sets. The **Cartesian product** of A and B is defined by:

$$A \times B := \{\{\{a\}, \{a, b\}\} \in \mathfrak{P}^2(A \cup B): a \in A \text{ and } b \in B\}.$$

We refer to an element, $\{\{a\}, \{a, b\}\}$, of a Cartesian product as a **couplet**. In the short term, we will use the visual artifice (a, b), with which you are familiar, as a stand-in for the notationally more cluttered $\{\{a\},\{a, b\}\}$. More on this later, when you will be hearing a lot about couplets.

Importantly, for our purposes, we note that the Cartesian product $\times$ distributes over the union operator $\cup$ as follows:

$$(A \cup B) \times (A \cup B) = (A \times A) \cup (A \times B) \cup (B \times A) \cup (B \times B).$$

If we leave that floating around in the middle of the page, you probably wont see what's happening, so we will employ a visual artifice to assist.

A good way to picture a Cartesian product is as a table, with each couplet occupying a cell in the table. So the Cartesian product of $A = \{2, 4, 6,\}$ with $B = \{1, 3, 5\}$ can be illustrated as follows:

$A \times B$		B		
		1	3	5
	2	(2,1)	(2,3)	(2,5)
A	4	(4,1)	(4,3)	(4,5)
	6	(6,1)	(6,3)	(6,5)

$A \times B = \{(2,1), (2,3), (2,5), (4,1), (4,3), (4,5), (6,1), (6,3), (6,5)\}.$

Incidentally, there is nothing to stop us from creating the Cartesian product of sets with different cardinalities, say $A = \{2, 4\}$, $B = \{1, 2, 3, 4\}$, as illustrated in the table below:

$A \times B$		B			
		1	2	3	4
A	2	(2,1)	(2,2)	(2,3)	(2,4)
	4	(4,1)	(4,2)	(4,3)	(4,4)

$A \times B = \{(2,1), (2,2), (2,3), (2,4), (4,1), (4,2), (4,3), (4,4)\}$.

So with $A = \{2, 4\}$ and $B = \{1, 4\}$, and let us now demonstrate that:

$$(A \cup B) \times (A \cup B) = (A \times A) \cup (A \times B) \cup (B \times A) \cup (B \times B).$$

$A \times A$		A	
		2	4
A	2	(2,2)	(2,4)
	4	(4,2)	(4,4)

$A \times A = \{(2,2), (2,4), (4,2), (4,4)\}$.

$A \times B$		B	
		1	4
A	2	(2,1)	(2,4)
	4	(4,1)	(4,4)

$A \times B = \{(2,1), (2,4), (4,1), (4,4)\}$.

Note, as we shall now illustrate, $A \times B$ is definitely not the same as $B \times A$ (i.e., $A \times B \neq B \times A$):

$B \times A$		A	
		2	4
B	1	(1,2)	(1,4)
	4	(4,2)	(4,4)

$B \times A = \{(1,2), (1,4), (4,2), (4,4)\}$.

See – definitely not the same as $A \times B$. So, on to $B \times B$:

$B \times B$		B	
		1	4
B	1	(1,1)	(1,4)
	4	(4,1)	(4,4)

$B \times B = \{(1,1), (1,4), (4,1), (4,4)\}$.

Stringing it all together,:

$(A \times A) \cup (A \times B) \cup (B \times A) \cup (B \times B) = \{(2,2), (2,4), (4,2), (4,4), (2,1),$
$(2,4),$

$\qquad\qquad\qquad (4,1), (4,4), (1,2), (1,4), (4,2), (4,4),$
$\qquad\qquad\qquad (1,1), (1,4), (4,1), (4,4)\}$

$\qquad\qquad = \{(1,1), (1,2), (1,4), (2,1), (2,2), (2,4),$
$\qquad\qquad\qquad (4,1), (4,2), (4,4)\}$

Now consider $(A \cup B) \times (A \cup B)$. Let $C = A \cup B$. Then $C = \{2,4\} \cup \{1,4\} = \{1, 2, 4\}$. So $C \times C$ is given by:

$C \times C$		C		
		1	2	4
C	1	(1,1)	(1,2)	(1,4)
	2	(2,1)	(2,2)	(2,4)
	4	(4,1)	(4,2)	(4,4)

So, $(A \cup B) \times (A \cup B) = (A \times A) \cup (A \times B) \cup (B \times A) \cup (B \times B)$ is true in this example, at the very least. And, of course, it is true in general.

We can thus assume that any set of couplets is a subset of a Cartesian product of the form $U \times U$, for an appropriately chosen U. Having noted that $A \times B \neq B \times A$, we need to state the relationship more formally, by writing:

$\qquad A \times B = B \times A$ if, and only if, $A = \emptyset$ or $B = \emptyset$ or $A = B$.

and also:

$\qquad A \times (B \times C) = (A \times B) \times C$ if, and only if, $A = \emptyset$ or $B = \emptyset$ or $C = \emptyset$.

If you remember the terms "commutative" and "associative," you can now confidently declare that the Cartesian product is neither commutative nor associative.

In reality, the expression $A \times B \times C$ is as dubious as the arithmetic expression[5] $6 - 3 - 7$ because there is no indication of which operator to apply first. Indeed, $A \times B \times C$ is notational nonsense unless $A = \varnothing$ or $B = \varnothing$ or $C = \varnothing$.

Heavy lifting

In Chapter 3, the arithmetic ascent from $\mathbb{Z}$ to $\mathfrak{P}(\mathbb{Z})$ served as our introduction to lifting, happily lending itself to basic notation:

"For $A, B \in \mathfrak{P}(\mathbb{Z})$, $A + B := \{a + b \in \mathbb{Z} : a \in A \text{ and } b \in B\}$",

and smooth talk:

"The sum of sets of integers is the set of the sums."

Recall, as we once told you, that the sequence:

$$\{\varnothing\}, \{\varnothing,\{\varnothing\}\}, \{\varnothing,\{\varnothing,\{\varnothing\}\}\}, \{\varnothing,\{\varnothing,\{\varnothing,\{\varnothing\}\}\}\},$$

is the genesis of the set of natural numbers:

$$\mathbb{N} := \{1, 2, 3,...\}.$$

True, we didn't prove that mathematically when we declared it so in Chapter 3, and neither are we going to prove it now. Nevertheless, please take our word for it.

Your parents (or maybe just your mother) knew about $\mathbb{N}$, because early on they (or she) turned $\mathbb{N}$ into the "counting" algebra $[\mathbb{N}; \{[+]\}]$ for you. No frills, such as zero or negation or even philosophical contemplation of the meaning of infinity; just counting, adding, commuting and associating with reckless abandon.

But, had your mom revealed to you the rudiments of set theory, then maybe, as a precocious mathematical child, together you would have explored the Cartesian product of this "counting algebra." You would surely have realized the sense in doing so, because your mom would doubtless have advised you that if your algebra seems somewhat limited, then you should ascend to a Cartesian product, like all precocious kids do.

So if you've had a mathematically deprived childhood, take the time now to ascend to the Cartesian product:

---

5   *Work it out. 6 – 3 – 7 could be 10 or it could be -4.*

$$\mathbb{N} \times \mathbb{N} = \{(a, b) \in \mathfrak{P}^2(\mathbb{N}): a, b \in \mathbb{N}\}$$

and amuse yourself by meandering around. Really, take a look:

N x N	N							
	...	...	...	...	...	...	...	...
	(1,7)	(2,7)	(3,7)	(4,7)	(5,7)	(6,7)	(7,7)	...
	(1,6)	(2,6)	(3,6)	(4,6)	(5,6)	(6,6)	(7,6)	...
	(1,5)	(2,5)	(3,5)	(4,5)	(5,5)	(6,5)	(7,5)	...
N	(1,4)	(2,4)	(3,4)	(4,4)	(5,4)	(6,4)	(7,4)	...
	(1,3)	(2,3)	(3,3)	(4,3)	(5,3)	(6,3)	(7,3)	...
	(1,2)	(2,2)	(3,2)	(4,2)	(5,2)	(6,2)	(7,2)	...
	(1,1)	(2,1)	(3,1)	(4,1)	(5,1)	(6,1)	(7,1)	...

Make a careful note of what we are about to do. Because we are about to demonstrate an ascension up the algebraic tower from $\mathbb{N}$ to $\mathbb{Z}$, using nothing more than the above Cartesian product – absolutely no smoke and no mirrors.

If you really take time studying this Cartesian product you might discover:

- Since we can add natural numbers, we can certainly lift + to $\mathbb{N} \times \mathbb{N}$, by defining the sum of couplets to be the couplet of sums: $(a, b) + (c, d) := (a + c, b + d)$: For example $(2, 3) + (4, 1) = (6, 4)$. So that gives us the algebra [$\mathbb{N} \times \mathbb{N}$; {+}]. And if you check, you will find that this lifted addition is pleasingly commutative and associative.

- Maybe the "diagonals" of our $\mathbb{N} \times \mathbb{N}$ array have caught your attention. If you follow the "main" diagonal traveling from bottom left to top right, it is populated by couplets of the form (x, x). The "higher" diagonal above, starting with $(1, 2)$ has couplets $(x, x+1)$. In general, the higher k-th diagonal has couplets $(x, x+k)$ and the lower k-th diagonal has couplets that are the transpose of $(x, x+k)$. i.e., $(x+k, x)$, where x ranges over $\mathbb{N}$.

- What does this portend? Are the higher diagonals positive in some sense and the lower diagonals negative in some sense? What does this say about the main diagonal? Could that be zero? And is the diagonal in fact a mirror? Corresponding higher and lower diagonals are indeed reflections of each other through the main diagonal.

- Does it look as though we have found the set of integers $\mathbb{Z}$ hiding playfully in **the diagonals of couplets of** $\mathbb{N} \times \mathbb{N}$?

Wait a minute. We are getting ahead of ourselves by designating these diagonals to be "entities" in their own right. Remember, $\mathbb{N} \times \mathbb{N}$ is a set of couplets, not a set of diagonals. The diagonals are merely a visual feature in a visual artifice and have no standing of themselves. Nevertheless, each diagonal is undeniably a set of couplets. And that means that each diagonal can be properly viewed as an element of the power set $\mathfrak{P}(\mathbb{N} \times \mathbb{N})$.

So watch closely, while we perform some genuine mathematical manipulations, and try not to get the impression that you're playing Three Card Monte with a street-wise card shark.

To maintain set-theoretic integrity, we must ascend to $\mathfrak{P}(\mathbb{N} \times \mathbb{N})$ and identify the elements of the subset, let us call it $\mathbb{D}$, of $\mathfrak{P}(\mathbb{N} \times \mathbb{N})$ whose elements are our diagonals, by numerals in bold font.

For example, **4** denotes:

$$\{(x, x+4)\} \in \mathbb{N} \times \mathbb{N} : x \in \mathbb{N}\}.$$

It follows that the transpose of **4**, let's represent it as $\overset{\leftrightarrow}{\mathbf{4}}$, denotes:

$$\{(x+4, x)\} \in \mathbb{N} \times \mathbb{N} : x \in \mathbb{N}\}.$$

We "invent" the symbol **0** to denote:

$$\{(x, x)\} \in \mathbb{N} \times \mathbb{N} : x \in \mathbb{N}\},$$

because this diagonal, and only this one, is not characterized by any element of $\mathbb{N}$.

We now lift "+" yet again, and restrict our attention to $\mathbb{D}$. This yields:

$$\mathbf{4} + \overset{\leftrightarrow}{\mathbf{4}} = \{(x, x+4) + (y+4, y) \in \mathbb{N} \times \mathbb{N} : x, y \in \mathbb{N}\}$$

$$= \{(x+y+4, x+4+y) \in \mathbb{N} \times \mathbb{N} : x, y \in \mathbb{N}\}$$

$$\subset \{(z, z) \in \mathbb{N} \times \mathbb{N} : z \in \mathbb{N}\}.$$

In other words, each possible sum of elements from components **4** and $\overset{\leftrightarrow}{\mathbf{4}}$ of our partition $\mathbb{D}$ of $\mathbb{N} \times \mathbb{N}$ is the component **0** of $\mathbb{D}$. So, by unanimous democratic vote, $\mathbf{4} + \overset{\leftrightarrow}{\mathbf{4}} := \mathbf{0}$.

Similarly:

$$0 + 4 = \{(x, x) + (y, y+4) \in \mathbb{N} \times \mathbb{N} : x, y \in \mathbb{N}\}$$

$$= \{(x+y, x+y+4) \in \mathbb{N} \times \mathbb{N} : x, y \in \mathbb{N}\}$$

$$\subset \{(z, z+4) \in \mathbb{N} \times \mathbb{N} : z \in \mathbb{N}\}.$$

Thus, each possible sum of elements from components **0** and **4** of our trusty partition $\mathbb{D}$ of $\mathbb{N} \times \mathbb{N}$ is the component **4** of $\mathbb{D}$. By overwhelming vote: $0 + 4 := 4$.

And, similarly:

$$4 + \overset{\leftrightarrow}{5} = \{(x, x+4) + (y+5, y) \in \mathbb{N} \times \mathbb{N} : x, y \in \mathbb{N}\}$$

$$= \{(x+y+5, x+4+y) \in \mathbb{N} \times \mathbb{N} : x, y \in \mathbb{N}\}$$

$$= \{((x+y+4) +1), x+4+y) \in \mathbb{N} \times \mathbb{N} : x, y \in \mathbb{N}\}$$

$$\subset \{(z+1, z) \in \mathbb{N} \times \mathbb{N} : z \in \mathbb{N}\} = \overset{\leftrightarrow}{1}.$$

And so, each possible sum of elements from component **4** and component $\overset{\leftrightarrow}{5}$ respectively, of our partition $\mathbb{D}$ of $\mathbb{N} \times \mathbb{N}$ is an element of the component $\overset{\leftrightarrow}{1}$ of $\mathbb{D}$. The votes are in, and the vote count indicates a landslide for:

$$4 + \overset{\leftrightarrow}{5} = \overset{\leftrightarrow}{1}.$$

Lo and behold, to your complete mathematical astonishment, it turns out that $\mathbb{D}$, an infinite set of infinite sets of couplets, together with lifted addition, lifted multiplication and lifted transposition is the authentic man-made integer algebra, which signs its checks in the following way:

$$[\mathbb{D}, \{[+, \mathbf{0}], [\times, 1], \{\leftrightarrow\}].$$

Now if you are going to quibble about the fact that we only demonstrated addition and not multiplication, just take our word that it's not sleight of hand that's in play here. The simple fact is that proving out multiplication in a similar manner to the above is very messy and thus, in time-honored fashion, we leave it as an exercise for the reader.

By the way, it was $\mathbb{D}$'s doppelganger, whose autograph is:

$$[\mathbb{Z}, \{[+, \mathbf{0}], [\times, 1], \{-\}],$$

that you encountered long ago in elementary school.

If you gaze at this famous signature and wonder to yourself how it could be that the "$\leftrightarrow$" transpose operator of $\mathbb{D}$ is the same as the "$-$" minus operator of $\mathbb{Z}$, we can only console you with the words: the truth is that they do exactly the same thing.

And if the ingrained learned-by-rote mantra "the negative of a negative is a positive" suddenly pops up in your consciousness at this point, you can append "and the transpose of a transpose is what you started with in the first place," knowing full well that they indicate exactly the same thing.

The fact is that your journey through school mathematics was a voyage into number systems of ever increasing computational richness and cardinality: from the naturals, to the integers, to the rationals, to the reals and then on to the complex numbers.

Most likely the impressive-looking relationships:

$$\mathbb{N} \subset \mathbb{Z} \subset \mathbb{Q} \subset \mathbb{R} \subset \mathbb{C}$$

appeared on the blackboard or the whiteboard at some point, as a summary of this progression. Sorry we have to point it out, but this impressive chain of relationships is not exactly kosher, or perhaps better put, it is notationally misleading.

For example, we know that the naturals are not a subset of the integers. As we have just demonstrated, it would be far more accurate to proclaim that there is a proper subset of the algebraically richer integers, consisting of those upper diagonals we constructed, that is algebraically equivalent to the natural numbers.

The upshot of this is not that we should curse our school mathematics teachers, who are anyway obliged to take one or two mathematical liberties in order to introduce us to foundational mathematical ideas. No. The real lesson is that our investment in a Cartesian product has yielded considerable algebraic profit.

And so it proceeds by numeric ascent. The rationals $\mathbb{Q}$ are derivative of $\mathbb{Z}$ and contain a clone of the integers. The reals $\mathbb{R}$ are derivative of $\mathbb{Q}$ and contain a clone of the rationals. The complex numbers $\mathbb{C}$ are derivative of $\mathbb{R}$ and contain a clone of the reals.

Here we can point at a whole numeric tower of algebraic ascension, supported by a very helpful scaffolding of Cartesian products. This series of lifts from one algebra to another was essential to the development of modern mathematics.

The same turns out to be true in the development of data algebras. There is a tower, there are Cartesian products and there is an elevator to help us ascend.

The Kuratowski Diversion

We would rather not involve you in the debates and deliberations of pure mathematicians, but we feel obliged to do so in the case of the "ordered pair." That terminology and the traditional visual artifice married to it, once out of the bag, fairly engenders unwelcome ambiguity.

Our story begins with Kazimierz Kuratowski and his set construct $\{\{a\},\{a,b\}\}$. Kuratowski offered this construct as the definition of an "ordered pair" in 1921, and it was generally accepted. So far, so good, in the sense that it simultaneously binds and distinguishes two distinct elements in one entity. Moreover, if the elements are not distinct (i.e., $a = b$), the issues of binding and distinguishing are moot, which is captured by:

$$\{\{a\}, \{a, a\}\} = \{\{a\}, \{a, a\}\} = \{\{a\}, \{a\}\} = \{\{a\}\}.$$

If this definition is not peculiar to a single element or a pair of distinct elements, then it begs for a generalization that maintains the set-theoretic integrity of the original while extending its reach up to as many as three distinct elements. The seemingly natural choice for an "ordered triple:"

$$(a, b, c) := \{\{a\}, \{a, b\}, \{a, b, c\}\},$$

appears promising: (a, b) and (a, b, c) are fellow members of $\mathfrak{P}^2(U)$, and this fellowship is shared by so-called "ordered n-tuples" that are defined in this manner. That's fine, at first blush, but it gives birth to anomalies. For example:

$$(a, a, a) = \{\{a\}, \{a, a\}, \{a, a, a\}\} = \{\{a\}\} = (a, a)$$

and:

$$(a, a, c) = \{\{a\}, \{a, a\} \{a, a, c\}\} = \{\{a\}, \{a, c\}\} = (a, c).$$

If you don't immediately see something amiss with these equations, then think of the Cartesian coordinates that we know and love. An "ordered triple" implies three axes, or dimensions if you prefer, and will not settle for two. Now let us pause while you work yourself into a frenzy at the mathematical atrocity, implicit in these equations. . . . . .

It is possible to deal with this and other problems by bending the rules – in effect by adding new axioms to the game of ZFC. But this is a price that we are not willing to pay for several reasons, one of which is that it would hinder our effort to treat with data in its full generality.

So what to do?

We shall ignore the begging, and never again in this text or in our description of data algebra use the term "ordered pair" or the visual artifice (a, b) in reference to the Kuratowski construct $\{\{a\}, \{a, b\}\}$. A couplet is a couplet is a couplet: a certain element of the second power set of some set of your choosing. By making no more of a couplet than that, we are able to make much more of couplets than that.

A visual artifice for the couplet

You probably realize from the above that the couplet plays a significant role in data algebra, otherwise why would we have troubled you with discussion of its definition? In addition to substituting the term "couplet" for the term "ordered pair," we shall now introduce some new mathematical notation.

The reason is this. In a chapter or two, we shall be obliged to discuss sets of couplets, sets of sets of couplets and so on. Imagine how that's going to look if we retain the braces of $\{\{a\}, \{a, b\}\}$ – a forest of braces all brawling with each other to define where we are and what mathematical object we are dealing with in our "towers of sometimes frightening height and complexity."

While we could have proceeded instead with the notation (a, b), we also judged this to be a pair of parentheses too far. In the end, we adopted the exponential form, deciding to use a^b in place of $\{\{a\}, \{a, b\}\}$. Our rationale was:

- a^b is less cluttered and more humane than the tortuous $\{\{a\}, \{a, b\}\}$.

- a^b visually captures the coalescence of a and b, while at the same time distinguishing them.

- a^b suggests a symmetric clone, b^a, which turns out to be important and which we will later refer to as the transpose of a^b.

- a^b and b^c almost beg to be composed; that is $a^b \circ b^c = a^c$, which indeed turns out to be a fertile binary operation in data algebras.

Now, at the very end of this chapter, it is time to introduce an important mathematical question:

What does it mean for a^b and x^y to represent the same couplet?

The facile answer, based on nothing more than the fact that a^b and x^y are just visual artifices, is that $a = x$ and $b = y$. Such a response is not only mathematically insufficient; to dyed-in-the-wool mathematicians, it is mathematically abhorrent. The point is that visual artifices do nothing more than represent mathematical objects, in this case, couplets, and if we are going to prove that $a^b = x^y$ implies that $a = x$ and $b = y$, we can only do so by appealing to actual mathematical entities:

$$\text{``}\{\{a\},\{a,b\}\} \text{ and } \{\{x\},\{x,y\}\}.\text{''}$$

To put it briefly, mathematics always dictates to visual artifices, whether they happen to be diagrams (such as Venn diagrams) or textual representations, and the vice is not versa.

Your mission, should you wish to accept it, is to prove that:

$$a^b = x^y \text{ if, and only if, } a = x, \ b = y.$$

First, a few words of warning: you are about to be beset by our only request in this book for you, the reader, to provide a real mathematical proof. It is elementary and short, and we believe in you; we believe you can pull it off. So please give it try. It is also subtle, so if you have trouble with it, move on, but don't give up. If necessary, come back to it from time to time (here's a hint: the proof involves nothing more than being careful with the axiom of extension). We have little doubt that you will eventually get it.

And it is important to get it, because algebra in general and data algebra in particular is, well, couplets galore!

Taking Stock

The following bullet points summarize this chapter:

- Relations (between elements of sets) rather than functions (of the form $y = f(x)$ as in numerical algebra) are the gold standard of data algebra.

- The Cartesian product of A and B is defined by:

$$A \times B := \{\{\{a\}, \{a, b\}\} \in \{\mathfrak{P}^2(A \cup B): a \in A \text{ and } b \in B\}.$$

- An element, $\{\{a\}, \{a, b\}\}$ of a Cartesian product is referred to as a **couplet**.

- The Cartesian product creates a set of couplets that can be illustrated in a tabular form. Couplets can represent Cartesian coordinates: points on a graph of a function, such as $y = x^2$.

- The Cartesian product "$\times$" is neither associative nor commutative. However, it distributes over the union operator " $\cup$ ", as follows:

$$(A \cup B) \times (A \cup B) = (A \times A) \cup (A \times B) \cup (B \times A) \cup (B \times B).$$

- Using the Cartesian product, $\mathbb{N} \times \mathbb{N}$, we can ascend from the set of natural numbers, $\mathbb{N}$, to the set of integers $\mathbb{Z}$. An involved explanation of how this works is provided, because we use a similar process of lifting to create algebras of data.

- We use the Kuratowski definition $A = \{\{a\}, \{a, b\}\}$ to define a **couplet**. In the sequel we discover that a "genuine" ordered pair is a certain set of two couplets and that an ordered n-tuple is a certain set of n couplets.

- Additionally, we introduce new notation (an exponential form) a^b to denote a couplet. This notation facilitates the eventual composition of couplets, and the so-called relations and clans they spawn, while keeping us from getting lost in forests of set braces.

Chapter 5: The Lion Sleeps Tonight

Go down deep enough into anything and you will find mathematics.

~ Dean Schlicter

—— ❊ ——

CONSIDER the statement:

In the jungle, the mighty jungle, the lion sleeps tonight.

You probably recognize this as a line from a decades-old pop song. However, is it data? Or to declare our focus of interest and be more precise: is it data as far as a computer is concerned? We cannot accurately answer such a question without first answering the more precise question:

What is the fundamental unit of data?

When we hear the word data it is more likely to conjure up the idea of information stored in a computer file or database than a sentence from a book or a line from a song. When we start to investigate this question, we quickly realize that it is not such an easy question to answer.

The algebraic necessity

Mathematically, we resolved the question of what data is, in a very general sense, by observing the fact that: data is a collection of "datums." This led us to set theory and thus defining the rules we agreed to play by – the axioms of ZFC set theory. But on its own, this does not provide us with a means of applying this to the data stored in a computer – and not just the data currently stored in computers, but any data that anyone may in the future decide to store in a computer.

It is an algebraic necessity that we start out with a sound definition of the fundamental unit of data which must apply even to data that may never be put into a computer. Consequently, it needs to be a very accurate definition and also one that is grounded in the world that we inhabit as human beings.

We might reason as follows: data defines our existence as living organisms, as human beings. Our means of understanding the world comes from information gathered by our six senses: touch, taste, smell, hearing, sight and the sense of motion and balance (proprioception). Our brain ingests a variety of such sensory impulses from our nervous system, sometimes combining one sense with another, and constructs a representation of the exterior world. We do this from before the moment of our birth, and we continue to do it until our last breath. So surely this representation of the world is a collection of data. Perhaps data is "the representation of anything we can perceive."

This is, in our view, a reasonable starting point, but it is not good enough for two reasons:

1.   It ignores our ability to conceptualize.

2.   It is subjective.

First we need to allow for the fact that we have the mental capacity to conceptualize, to represent concepts and even physical things that do not yet exist in reality. This too might be regarded as data, especially as such mental models may eventually become real-world data. The whole of mathematics is a collection of such mental models that have been committed to paper and shared extensively. Indeed, we could say the same about many conceptual models that inhabit the broad field of human intellectual endeavor.

The second point is much more troublesome for our initial stab at a definition of data. It is a philosophical and psychological cliché that when two people claim to see the color red, there is no guarantee that their experience is identical. And the existence of color blindness demonstrates that, in some circumstances at least, it definitely is not. Even if we could place electronic sensors in the human brain to record and identify signals emerging from the nervous system and were thus able to prove that most people receive identical signals at the interface to their brain, we cannot know for sure that their interpretation of them will be the same.

This means that we must go to objective sources, directly to the physical activity of man, in order to define the what data is.

Human information

First consider text. For millennia humanity has communicated via the written word, chiselled into stone, or painted onto walls or written on papyrus or vellum or paper. The goal was to present data for others to read, whether they were contemporaries or future generations. We can thus examine data from this perspective.

Text, whether composed of alphabetic letters or glyphs, consists of words arranged into sentences. Numbers are the same and can be viewed as words of a kind created from a special alphabet (digits). So the basic building blocks of text are a set of symbols.

The first question then is whether these building blocks can be thought of as data. Although the symbols do not necessarily have explicit meaning in isolation – sometimes they do, sometimes they don't – when used together they impart meaning. So we suspect that they may constitute data, since they are components of data. However, with words, the units of text that the symbols create, there is much less doubt. Words have meaning.

For the sake of exactness we need to consider all arrangements of symbols that have meaning when taken together. So this includes numeric units of meaning, such as 3.14159, as well as verbal units of meaning, such as "lion" and "jungle." Words have meaning and numbers have meaning. As we construct sentences we create more sophisticated and more detailed meaning.

So does this provide a basis for us to define what we mean by data?

It doesn't.

The demands of objectivity

The problem is this. Words and numbers constructed in this fashion require a human being to decipher their meaning. If you ask several people what the word "lion" means, you will quickly realize that there is no reliable consensus as to its meaning.

The answer you get from any individual will vary according to their experience which, in respect to the word "lion," consists of a collection of memories, some of which they will associate to when you ask the question. Their answer will almost certainly not be identical with a dictionary definition.

Look the word up in the *New Oxford American Dictionary*, for example, and it will suggest that a lion is "a large tawny-colored cat that lives in prides, found in Africa and northwestern India. The male has a flowing shaggy mane and takes little part in hunting, which is done cooperatively by the females."

The fundamental problem here is that the meaning people infer from "lion" varies according to their individual experience. Written down, the word "lion" is simply four alphabetic symbols in a row, pronounced in a particular fashion when verbalized. Consult a dictionary, and you will encounter a reasonable definition perhaps, yet even there you will not encounter agreement between dictionaries as to what should be included in the definition.

For contrast, consider the *Merriam-Webster* definition of "lion" (from merriam-webster.com) which states that a lion is "a large heavily built social cat (Panthera leo) of open or rocky areas chiefly of sub-Saharan Africa though once widely distributed throughout Africa and southern Asia that has a tawny body with a tufted tail and a shaggy blackish or dark brown mane in the male."

And it is only fair to point out that *New Oxford American Dictionary* also adds the following alternative meanings or usages:

- (the Lion) the zodiacal sign or constellation Leo.

- a brave or strong person.

- an influential or celebrated person: a literary lion.

- the lion as an emblem (e.g., of English or Scottish royalty) or as a charge in heraldry.

The Merriam-Webster alternative meanings or usages are not identical. They are:

- any of several large wildcats; especially: cougar.

- LEO (this, incidentally, is a link to the definition of LEO the constellation).

- a person felt to resemble a lion (as in courage or ferocity).

- a person of outstanding interest or importance <a literary lion>.

When the word "lion" comes up in conversation or is written on a page, any one of those meanings – some of which are metaphorical – may be intended. When we read the word, usually without any conscious effort, we select the intended meaning. We deduce its meaning from the context. So when we read the words:

In the jungle, the mighty jungle, the lion sleeps tonight,

we do not suppose that the constellation Leo, or some literary lion, or some brave or strong individual has nodded off in the jungle. We are pretty certain that Panthera leo himself has decided to call it a night.

This is all very well for human beings, but it creates a problem for computers. Computers can only make the estimations of meaning we have described here if they have similar mechanisms for determining meaning. And usually they do not have any such capabilities, although it is worth noting here that their semantic and reasoning skills are currently improving by leaps and bounds.

Consider something quite simple, such as an instant message from one person to another, as follows: "meet u after work 4 drink." Even someone who has never previously encountered the abbreviation "u" for "you" and "4" for "for" would be able to work out the meaning. They would deduce it. A computer however, would be unlikely to have a basis for making such a deduction. Computers require data to be rigorously defined, to be objective. And so does an algebra of data.

The basic unit of data

We already know a great deal about what works as data from the way we use computers. We do not put information into a computer without representing it in a way that the computer can decipher and giving it sufficient context to allow us to do something useful with it. We know what we have found to be practical.

At the base level, we store information as encoded bits. The natural unit of computer data is called a "word." A specific definition for a word is required by the computer processor, since the instructions that the processor can execute only process data in units of words. In practice word sizes of 8, 16, 24, 32, 48 and 64 bits have been tried at various times and nowadays most computers use 32 bit or 64 bit words.

In general, although not always, the word determines the address size for addressing data held in memory. Machine instructions, the basic instructions executed by the processor, will refer to (point at) addresses of data held, and the data is fetched into the processor so that the instructions can be executed.

At the very basic level, ignoring a good deal of the nuances of computer architecture, we can think of computer programs as a series of instructions that specify addresses of data. When these instructions are executed by the processor, data is fetched, processed and the result is placed somewhere in memory. The instruction set that the processor executes is itself a set of binary codes, and the data contained in the words stored in computer memory is also binary. At this level, we could say that there are just two kinds of data: instructions stored in words and data stored in words.

At a slightly higher level the computer program distinguishes between different types of data. Exactly which data types are accepted by a computer program depends on the programming language. The following comprise the most common data types:

- **The Boolean data type**. This either has the value "True" (usually represented by a single bit with value 1) or "False" (represented by a single bit with value 0).

- **The integer data type**. This is an integer. Depending upon computer language and architecture there can be short integers, long integers and even long long integers, and they may be signed (meaning that they include the numeric sign, + or -). An unsigned long long integer can take any value from 0 up to 18,446,744,073,709,551,615.

- **The floating point data type**. This numeric form allows computers to represent and compute real numbers. They are called floating point because the number's decimal point can "float," so that very large numbers or very small numbers can be represented even if the representation is approximate to some degree. The number has two parts called the significand (an integer) and the exponent (also an integer). The number $3.14159 = 314159 \times 10^{-5}$ is thus stored as 314159 (the significand) along with -5 (the exponent). The limit on the size of the number that can be represented depends on the number of bits used for the significand and the exponent. The "single" floating point usually occupies 32 bits with 23 for the significand and 8 for the exponent and 1 for the sign. The double is 64 bits (52, 11, 1), and the quad 128 bits (112, 15, 1).

- **The char data type**. This is one character and is best thought of as a single keyboard character since the value usually corresponds to something that is visible on the computer keyboard.

That's all the basic data types, although most programming languages offer a richer capability than that, as do databases. The char naturally extends to become a string of characters, usually with a termination character to denote the end of the string. Date and date-time may have a dedicated data type. Arrays and lists are often supported by programming languages. There are no explicit data types for images or vector graphics or sound or video at the language level, but there are agreed standards (often several) for the representation of such data.

The point here is that values are not stored within a computer without there being some indication of what that specific value means. No matter where data values are stored, it is impossible to process them meaningfully without knowing what they represent.

In formulating the algebra of data, the couplet was chosen as the fundamental unit of data. Our discussion of the basic data types of a computer indicates that there is strong circumstantial evidence for suspecting that a couplet is the correct set-theoretical form. The bare metal computer is "objective" in the sense that it has no in-built mechanism for processing data that is not described to it explicitly. Its processing is couplet-based.

So a couplet appears to be a respectable starting point, as a "unit of meaning." If we put two values together in a couplet, for example, 9^{lions} (yes, we are using use the couplet notation from the previous chapter) we can think of each as qualifying or providing context to the other. That set might, for example, be a count of lions in a particular location. It is possible, of course, to randomly pick any two units of meaning and create a couplet, say, $telephone^{liquid}$, where the two values cannot meaningfully provide context for each other (except in a Salvador Dali painting). This just indicates that some couplets are not particularly useful.

At the logical level

We can think of the computer as having two levels: a logical level where it represents data in a way that is convenient for people to use directly and a physical level where it stores data in a form that is convenient for its own needs. Any algebra of data worth our attention must be able to represent and operate on data at both levels and pass data from one level to the other.

So let us now consider the statement:

In the jungle, the mighty jungle, the lion sleeps tonight.

Is this data?

This certainly seems to be data that people process naturally. And without defining precisely what we mean by couplet mathematically, we can certainly represent this statement as a couplet in the following way:

In the jungle, the mighty jungle, the lion sleeps tonight.^{lyrics}.

By doing this we have designated the statement as "lyrics." That's one way of thinking about it. However, you will no doubt have noted that this is indeed a sentence, and as such, we could, if we wished, express each word of it as a couplet grammatically. So, it may occur to you that we can represent it as the couplet S^s, where S = {lots of couplets} and s = lyrics:

$\{In^{preposition},\ the^{definite\text{-}article},\ jungle^{noun},\ ,^{comma},\ the^{definite\text{-}article},$
$mighty^{adjective},\ jungle^{noun},\ ,^{comma},\ the^{definite\text{-}article},\ lion^{noun},$
$sleeps^{verb},\ tonight^{adverb}\}^{lyrics}.$

By the way: yes indeed, a set can be an element of a couplet (sets can be elements of sets). If we remove repetitive elements, we get:

$\{In^{preposition},\ the^{definite\text{-}article},\ jungle^{noun},\ ,^{comma},\ mighty^{adjective},$
$lion^{noun},\ sleeps^{verb},\ tonight^{adverb}\}^{lyrics}.$

If we wanted to preserve all the words in the correct order, we'd have to include order preserving information. This would give us:

$\{\{In^{preposition}\}^1,\ \{the^{definite\text{-}article}\}^2,\ jungle^{noun}\}^3,\ \{,^{comma}\}^4,\ \{the^{definite\text{-}article}\}^5,$
$\{mighty^{adjective}\}^6,\ \{jungle^{noun}\}^7,\ \{,^{comma}\}^8,\ \{the^{definite\text{-}article}\}^9,\ \{lion^{noun}\}^{10},$
$\{sleeps^{verb}\}^{11},\ \{tonight^{adverb}\}^{12}\}^{lyrics}.$

And this, too, is a couplet. But it is couplets within couplets within a set that's a component of a couplet. We imposed order on the set so that we did not lose meaning. And, of course, we can descend to a further level of detail if we wish to break up the individual words into their constituent letters. We might decide to transform every string that is noun and is recorded as the couplet $\{string^{noun}\}$ into such a set of compound couplets, as follows:

$\{\{j^{letter}\}^1,\ \{u^{letter}\}^2,\ \{n^{letter}\}^3,\ \{g^{letter}\}^4,\ \{l^{letter}\}^5,\ \{e^{letter}\}^6\}^{noun}.$

Thus, if we were so inclined, we could express the statement as a very complex structure, breaking it down into every single symbol it contains. And, by the way, to store this simple sentence in a computer we do indeed break it down to that level in some way – even though we currently do not do so in a respectable algebraic manner.

We are obliged to point out that we have made some assumptions about this sentence. We have, for example, assumed that the statement itself is just lyrics and not part of a wider context. Perhaps, in order to express the full meaning of what it represents, the data needs to be married with the musical notes that accompany it. There is nothing explicit about the original statement that insists it is not music. So, perhaps the data we should be storing is this:

Or perhaps we should only be storing a sound recording of that line of the song. And since it is a song, and thus covered by copyright, perhaps we need to investigate whether we have the legal right to store this in any form whatever. Perhaps we do not.

Taking another tack, we are actually presuming that this data refers to the lyrics of a song. Perhaps this is not the case. Perhaps it really is just a factual statement intended to convey meaning, in which case we can represent it as a couplet in the following way:

Tonight the lion sleeps in the mighty jungle[fact].

Notice here that the order of words may no longer be so important, since several different versions of this sentence will be comprehensible to any reader (The lion sleeps tonight in the mighty jungle; The lion sleeps in the mighty jungle tonight, and so on).

But sadly, with this example, we are unlikely to want to classify it as fact, since lions are never found in jungles. Their habitat is the Savannah, where they hunt buffalo, zebras and wildebeest, and where, sometime after nightfall, they go to sleep.

Taking yet another tack, we note that we are making definite assumptions about the individual to whom the data may be served up. The data consumer might be illiterate, or even if literate, the English language may be completely foreign to him or her. It is even possible that the data consumer does not recognize the English alphabet, and thus if they encounter the sentence, they may not even be certain that it is genuine text in some foreign language.

Consider this:

է շունգլիներում հզոր շունգլիներում առյուծը քնում երեկո

Do you know whether this is genuine text or not? Do you even know whether you should read it from right to left or left to right? You probably don't. In fact, it should be read from left to right. It is Armenian script. Roughly translated into English, it reads:

In the jungle, the mighty jungle, the lion sleeps tonight.

The point is that information may have structure in the sense that, if served up to the right person, meaning can be gleaned from the information, but the computer itself does not explicitly know this unless it is given some indication that this is so.

In connection with this, we note that the term "unstructured data" is frequently used in IT. However, it is poorly chosen terminology because it implies, when interpreted literally, that there can be some data which has no structure whatsoever. In fact, the meaning of "unstructured," according to Wikipedia, is:

Unstructured data (or unstructured information) refers to information that either does not have a pre-defined data model or is not organized in a pre-defined manner. Unstructured information is typically text-heavy, but may contain data such as dates, numbers, and facts as well.

In other words "unstructured data" can have implicit structure but has no explicit structure. Our example is, by this definition, "unstructured" data. It is certainly the case that we could feed the data into a computer and have it recognize the input as text. However, it could not know for sure that it was text, it could only know that it matched the pattern of text. Nevertheless, if that is what we wanted the computer to do, then we would have applied a function, let us say, f with the following effect:

$$f(\textit{In the jungle, the mighty jungle, the lion sleeps tonight.}^{input}) :=$$
$$\textit{In the jungle, the mighty jungle, the lion sleeps tonight.}^{text}.$$

It is difficult to imagine any data that has no structure. Even if we think of the various radio telescopes involved in the SETI project, which listen to and record radio waves from everywhere in the sky, hoping to find evidence of extraterrestrial life, we cannot know for sure that what they have recorded is simply random noise. All we know is that no-one who has analyzed such signals has yet been able to detect structure in them.

For all we know, some of these signals have a structure that we have failed to recognize. Perhaps, buried somewhere in that extensive record of extra-terrestrial radio signals, is the plaintive sound of some alien voice singing *The Lion Sleeps Tonight*.

The subjectivity of man

Hopefully, we have said enough to demonstrate that although the world of man has aspects of objectivity – people tend to understand each other when they speak the same language – the world of the computer is more demanding. In the world of man, our simple example, *In the jungle, the mighty jungle, the lion sleeps tonight.*, qualifies as data among a subgroup of humanity that recognizes the English language. However, it only does so because we have inner intellectual mechanisms which can parse the data and categorize it accordingly.

The computer, completely lacking any such mechanism, needs to have the context of that string of characters specified. As such, we have no option but to declare that *In the jungle, the mighty jungle, the lion sleeps tonight.* is not data as far as a computer is concerned. However, if we turn it into a couplet, as follows:

In the jungle, the mighty jungle, the lion sleeps tonight.^{string}.

then the computer can treat it as data at the physical level. We can write programs, which manipulate strings and feed such data to them. If we want to define this within the computer in a way that might be useful to people, we may prefer that couplet to be defined as follows:

In the jungle, the mighty jungle, the lion sleeps tonight.^{lyric}.

or possibly, we might be satisfied with:

In the jungle, the mighty jungle, the lion sleeps tonight.^{text}.

The Algebra of Data

All of this suggests that the objective unit of data genuinely is the couplet, and hence couplets are an excellent place to begin in an effort to construct an algebra of data.

Taking Stock

This chapter dealt with the question: What is data? The following bullet points summarize its content:

- The inner (psychological) world of man is demonstrably subjective.

- The outer consensus world of man is also demonstrably subjective. Where it is able to achieve objectivity, person to person, it does so through context.

- At the basic physical level within a computer, data is represented using couplets. One part of the couplet specifies value, the other part specifies usage.

- Couplets can also be used for representing data at the logical level, in terms that people can understand.

- The algebra of data takes the couplet to be the fundamental unit of data.

Chapter 6: Yin and Yang

How can it be that mathematics, a product of human thought independent of experience, is so admirably adapted to the objects of reality?

~ Albert Einstein

———⚬◦∞◦⚬———

IF YOU HAVE FOLLOWED our trail of bread crumbs into the not-so-fearsome forest of data algebra, you will now be familiar with power sets. You will have understood their importance: that you can build a tower from such sets and that you can climb energetically from one level to another, lifting existing operators and creating new operators as you go, as dictated by opportunity and need. You will have refamiliarized yourself with unions, intersections, complementation and possibly even with Cartesian products, if you previously met them at some point of your education.

You have just been introduced to the idea that the couplet, a^b, is the true basic unit of data and, to carve this idea in stone, you've been led through a discussion of what human beings happily accept as data and what computers are constrained to accept as data. The main point to emerge from this is that computers are an aggregation of dumb electronic circuits that need to be precisely informed about data, while human beings, blessed with an associative memory and some ability to reason, can manage with less precision. As the algebra of data is built for the computer's world, it conforms to the limits of the computer's world.

Despite the fact that we just spent a chapter discussing the validity and importance of the couplet, we have more to say. Basic units are indeed important...

A little more about couplets

We can commence with the fact that all civilizations develop or inherit units of measure. Such units, whether they apply to time, length, angle, weight, liquid volume or money, become the foundation of many activities, and without these units of measure, the activities would not be possible. If, for example, you were an apothecary in Great Britain in any century prior to the 20th, you would have known that there are: 20 grains to a scruple, 3 scruples to a drachm, 8 drachms to an ounce and 12 ounces to a pound.

You would also have had a set of weights that allowed you to weigh out such quantities, so that you could properly cost raw materials and price the various remedies you sold. Nowadays, of course, we have streamlined units based on decimal multiples. With only a few exceptions, we measure distances with meters, weights with grams, liquid volume with liters and so on.

So when we encounter a number, it is usually accompanied by a unit, suggesting very strongly that the numbers of the real world are really couplets of the form:

$$n^{unit}.$$

This is also true of the so-called pure numbers of theoretical mathematics. Those devil-may-care mathematicians have been happily writing numbers such as 1, 1.618, 3.1412 and so on as if they weren't couplets, and they are. And so are all those variables, x, y, and z, that star in their magic algebraic equations; couplets, every one of them. Depending on what they are, we can formally write them as one of the following:

$$n^{natural}, i^{integer}, q^{rational}, r^{real}, c^{complex}.$$

But of course, mathematicians finesse all of that confidently and safely assuming that the lowliest number they deal with lives somewhere in that notationally misleading tower:

$$\mathbb{N} \subset \mathbb{Z} \subset \mathbb{Q} \subset \mathbb{R} \subset \mathbb{C}.$$

Couplets ($\mathcal{G}$)

Couplets ($\mathcal{G}$) is an algebra. Once we have told you all there is to know about it, you may conclude that it is meager and unexciting; a couple of operators and a collection of data to get algebraic with – so what? However, this is a rags to riches story, and the rags of the couplet algebra will look more and more like royal robes as we climb the algebraic tower. So wait for that with bated breath.

In the meantime, as we're talking about Couplets ($\mathcal{G}$), let's talk about the genesis set. The couplets have to come from somewhere, and the genesis set $\mathcal{G}$ is the bag of things from whence they emerge. So what's in the bag? Officially and pragmatically, it's the "local" universe of "stuff," which is sufficient to generate all the couplets we wish to consider and manipulate. And just to emphasize the point, it is finite, even if large.

To get the couplets we want, we manufacture them using a Cartesian product. Imagine if there were only three things in the bag $\mathcal{G}$: red, yellow, leather. Not much content, but we would still be able to construct 9 couplets as follows:

$\mathcal{G} \times \mathcal{G}$		$\mathcal{G}$		
		leather	red	yellow
$\mathcal{G}$	leather	$leather^{leather}$	$leather^{red}$	$leather^{yellow}$
	red	$red^{leather}$	red^{red}	red^{yellow}
	yellow	$yellow^{leather}$	$yellow^{red}$	$yellow^{yellow}$

And while we might not be able to do much with those couplets, we can at least list the well-known tongue twister:

$$red^{leather}, yellow^{leather}, red^{leather}, yellow^{leather}.$$

If only we had some functions...

The functions: yin and yang

We will, at times, want to get our hands on the one or the other of the two components of a couplet, so we need a pair of functions to extract them. This pair of functions are named **yin** and **yang**, terms borrowed from Chinese philosophy, where **yin** and **yang** are complements rather than direct opposites. These words were not chosen by accident, but selected because they carry no implication of order.

So let there be two sets, A and B, with:

$$a^b = \{\{a\},\{a, b\}\} \in A \times B.$$

Then:

$$yin\,(a^b) = a \text{ and } yang\,(a^b) = b.$$

And now, without missing a beat, let's talk about the transposition operator. It should come as no surprise that we can transpose couplets. Recall back in Chapter 4, when we climbed the numeric tower, you will remember transposition, $\leftrightarrow$, to be a unary operator, that "switches things around." This is how it works on couplets:

$$\overleftrightarrow{b^a} = a^b$$

So, as you can immediately deduce, the transpose of $red^{leather}$ is $leather^{red}$, that the transpose of a transpose is, as always, what you started with and the transpose of a couplet like $yellow^{yellow}$ is just the same, $yellow^{yellow}$.

If we now introduce **yin** and **yang**, and we can declare that:

$$yin\ (red^{leather}) = red,$$

$$yang\ (red^{leather}) = leather.$$

And if you transpose the $red^{leather}$ to become $leather^{red}$ then the **yin** and **yang** results switch around so that:

$$yin\ (leather^{red}) = leather,$$

$$yang\ (leather^{red}) = red.$$

As you will realize later, given the transpose operator and the **yin** and **yang** functions, we can get at any part of a couplet whenever we want.

Composition

We promised you two operators for our modest Couplets ($\mathcal{G}$) algebra, so let's deliver on this promise. The second operator is a binary operator, and it's useful, and it's called **composition**. When the composition operator, "∘" composes two couplets, it behaves like this:

$$a^b \circ b^c = a^c.$$

To precisely define composition, we express it as follows:

$$a^b \circ d^c = a^c \text{ if, and only if, } b = d.$$

You may be wondering, "Well, what the hell happens if b is not equal to d?" The answer is, "In that case, there is no defined result." It's what is called a partial operator; sometimes it gives a meaningful answer, and sometimes it just doesn't.

Consider this for example:

$$red^{leather} \circ leather^{yellow} = red^{yellow}.$$

whereas:

$$red^{leather} \circ red^{yellow} \text{ is undefined.}$$

That's the downside of composition in Couplets ($\mathcal{G}$) but there is also an upside. Composition is associative!!

OK, we realize that you are not going to leap out of your chair and cheer just because of that, but mathematically, it's a real positive. Composition is associative:

$$a^b \circ (b^c \circ c^d) = a^b \circ b^d = a^d = a^c \circ c^d = (a^b \circ b^c) \circ c^d,$$

which is definitely associativity.

Or we can write:

$$red^{leather} \circ leather^{yellow} \circ yellow^{red}$$

and it doesn't matter which composition we work out first; we'll end up with the answer red^{red}.

The transposition of the composition

Now for something unexpected. We'd really like it if the transposition of a composition was the composition of the transpositions. Apart from anything else, we'd be able to blurt out our *flarn-of-the-clarp* mantra. But it isn't. The transposition of the composition is actually the composition of the transposition in the reverse order!

$$\overleftarrow{a^b \circ b^c} = \overleftrightarrow{a^c} = c^a = c^b \circ b^a = \overleftrightarrow{b^c} \circ \overleftrightarrow{a^b}.$$

And, as long as you remember this curious backflip when you're wrestling with data, it will serve you well.

Composition within the computer

Composition may not seem particularly useful at first, but it is the kind of operation that happens within a computer time and again – it just doesn't happen mathematically (yet). We know for example that a data item, when defined in a computer program, may be defined as:

$$Int^{age}.$$

Thus, a metadata tag (age) is bound to a data type (Int). The choice of data type was made most likely by the programmer. But when the computer comes to process the data item, the processor core will specify a memory address where that Int is to be found. Let us imagine that this address is the 64 bit value EG1A037F. At that level of activity, the computer is not concerned with what the integer residing at that location represents. It will simply think in terms of:

$$EG1A037F^{Int}$$

when it fetches the data for processing. And if it uses the data in some calculation or even simply moves it somewhere else, it will be handling the integer's value. Let's say the value is 35. Then the computer will have carried out the composition:

$$35^{EG1A037F} \circ EG1A037F^{Int} = 35^{Int}.$$

And if it later uses the metadata tag, age, to describe the data when it displays it in a window, it may carry out the composition:

$$35^{Int} \circ Int^{age} = 35^{age}.$$

In general, we can express an item of computer data with the couplet:

$$((value^{location})^{type})^{tag}.$$

As we have not shown you a compound couplet like that before, it may not have occurred to you that we could create such a thing. However, the truth is we can include anything we choose in a genesis set, including a compound couplet such as $(value^{location})^{type}$. The **yin** of a couplet can be a couplet and so can the **yang**. And we can apply our oriental functions to get at the bits we want. So, for example:

$$yin\ (yin\ ((value^{location})^{type})^{tag}) = value^{location},$$

$$yin\ (yin\ (yin\ ((value^{location})^{type})^{tag})) = value,$$

$$yang\ ((value^{location})^{type})^{tag} = tag,$$

$$yang\ (yin\ ((value^{location})^{type})^{tag}) = type,$$

$$yang\ (yin\ (yin\ ((value^{location})^{type})^{tag})) = location.$$

Also, introducing some composition:

$$yin\ (value^{location} \circ location^{type}) = value,$$

$$yang\ (value^{location} \circ location^{type}) = type.$$

And by the way, we could have driven you half insane by throwing the transpose operator into the various equations shown above and making the data do somersaults and backflips, but we judged it unwise.

Further extensions to couplets

There isn't really any limit to how complex a couplet can become, because a couplet can always be the *yin* of a couplet, or the *yin* of the *yin* of the *yin* an element of a couplet.

Consider a lion, whose name is Ralph. The following couplet is quite valid, and it's meaningful:

$$((((((Ralph^{lion})^{panthera})^{felidae})^{carnivora})^{mammalia})^{chordata})^{animalia}.$$

There is quite a lot of knowledge stored implicitly in this couplet, since, aside from the fact that Ralph happens to be of the species "lion," that species is part of the genus "panthera," which in turn is of the family "felidae," which is of the order "carnivora" within the class "mammalia" inside the phylum "chordata," all sitting inside the kingdom "animalia."

This demonstrates the possibility of adding qualification after qualification to a couplet. However, if you prefer, you can completely reverse it from beginning to end and represent the same data as:

$$((((((animalia^{chordata})^{mammalia})^{carnivora})^{felidae})^{panthera})^{lion})^{Ralph}.$$

As already noted once or twice, a couplet does not imply any ordering of any kind. So both representations are fine, although one may be preferable to the other depending on how you intend to use the data. If you examine either of them, you may think that some additional information needs to be explicitly included such as words like "species," "genus," etc. We could capture that information using a set of couplets. One way to do that would be:

$$\{Ralph^{lion}, lion^{panthera}, panthera^{felidae}, felidae^{carnivora}, carnivora^{mammalia},$$
$$mammalia^{chordata}, chordata^{animalia}, species^{lion}, genus^{panthera}, family^{felidae},$$
$$order^{carnivora}, class^{mammalia}, phylum^{chordata}, kingdom^{animalia}\}.$$

But we cannot do that in our lowly Couplets($\mathcal{G}$) algebra; we will have to lift it up the tower to do that. However before we do, let's declare a signature for Couplets($\mathcal{G}$).

The signature of Couplets($\mathcal{G}$)

While couplets are based on a **genesis set** $\mathcal{G}$, the **ground set** from which the couplets emerge is the Cartesian product of $\mathcal{G}$ with itself, $\mathcal{G} \times \mathcal{G}$. We can thus propose that the signature of Couplets ($\mathcal{G}$) is:

$$[\mathcal{G} \times \mathcal{G}, \{\circ\}, \{\leftrightarrow\}].$$

In plain English, this is saying that the algebra Couplets ($\mathcal{G}$) involves couplets created by the Cartesian product of $\mathcal{G}$ with itself and includes two operators, the partial binary composition operator and the unary transposition operator. And, by the way, that binary composition operator is not just partial, it also has no identity.

We have seen something similar to this with the natural numbers $\mathbb{N}$ and its associated algebra being limited but its capability being increased by lifting it to the integers $\mathbb{Z}$. So witness what happens when we ascend the algebraic tower by lifting Couplets ($\mathcal{G}$) to Relations($\mathcal{G}$).

Relations($\mathcal{G}$)

First things first; a relation is a set of couplets. If we have some couplets, say, $Thor^{hammer}$, $Hulk^{anger}$, $Iron\ Man^{suit}$, and we wish to relate them to each other in a set, we can wrap them up in braces like this:

$$\{Thor^{hammer},\ Hulk^{anger},\ Iron\ Man^{suit}\},$$

and we have created a relation. Couplets, as you well know (we hope) are created by Cartesian products. So the existence of those couplets implies the existence of corresponding sets, say:

A set of comic book superheroes, $\{Thor,\ Hulk,\ Iron\ Man,...\} \subset \mathcal{G}$.
A set of comic book superhero powers, $\{anger,\ suit,\ hammer,...\} \subset \mathcal{G}$.
A Cartesian product, $\mathcal{G} \times \mathcal{G} = \{Thor^{hammer},\ Hulk^{anger},\ Iron\ Man^{suit},...\}$.

Relations($\mathcal{G}$) is the algebra of all sets R, where $R \in \mathfrak{P}(\mathcal{G} \times \mathcal{G})$. In particular,

$$\{Thor^{hammer},\ Hulk^{anger},\ Iron\ Man^{suit}\} \in \mathfrak{P}(\mathcal{G} \times \mathcal{G}).$$

Incidentally, you may we wondering why we chose to call a set of couplets a relation. Perhaps you suspect the word has been chosen at random, and we could just as easily have called it a quark or a boson. Or maybe you believe we are intent on irritating aficionados of relational theory who, when asked "What is a relation?" will often reply, "It's a table, of course."

In truth, the word "relation" wasn't so much chosen as it chose itself. A set of couplets was created for some reason that justified the assembly of its elements, and that reason, no matter what it was, is a relation. Even if you decide to randomly pick a set of apparently unrelated couplets, it is a relation, and the relation that holds the elements together is "the set of apparently unrelated couplets I chose to pick."

Come yin, come yang

So now that we have a new tier of the tower to explore, let's see what we can do with our beloved functions *yin* and *yang*. Luckily, there is an easy and obvious way to apply them to relations. Let's show and tell. The *yin* of a relation is the set of the *yin*s of the elements of the relation, and naturally, the *yang* of a relation is the set of the *yang*s of the elements of the relation. So, consider:

$$R = \{Pinochio^{puppet},\ Miss\ Piggy^{muppet},\ Popeye^{cartoon}\}.$$

Then:

$$yin\ R = \{Pinochio,\ Miss\ Piggy,\ Popeye\}$$

is the yin set of R and:

$$yang\ R = \{puppet,\ muppet,\ cartoon\}$$

is the yang set of R.

Also, conveniently and with little intellectual effort, we can lift the operation of transposition into the algebra of relations:

$$\overset{\leftrightarrow}{R} = \{\ b^a \in B \times A:\ \overset{\leftrightarrow}{b^a} \in A \times B\}.$$

In English: *The transpose of the relation R is the relation whose elements are the transposes of the elements of R.* It's one of those *flarn-of-the-clarp* situations. A simple cartoon example will make this even clearer:

$$T = \{Bugs^{Bunny},\ Daffy^{Duck},\ Wiley^{Coyote}\},$$

$$\overset{\leftrightarrow}{T} = \{Bunny^{Bugs},\ Duck^{Daffy},\ Coyote^{Wiley}\}.$$

Now that we have the defined the transpose of a relation, we can introduce some new and useful yin-yang terminology. So let's introduce the important concepts of yin-functional and yang-functional relations:

If $a^b \in R$ and $a^c \in R$ imply that $b = c$ then R is said to be **yin-functional**.

As is visually apparent, our relation T is yin-functional.

We can concoct a similar definition for the meaning of yang-functional:

R is **yang-functional** if $\overset{\leftrightarrow}{R}$ is yin-functional.

Take a look at $\overset{\leftrightarrow}{T}$ and you see also that T is yang-functional. And just to be sure that you've got this, we'll consider a different cartoon relation:

$$C = \{Donald^{Duck},\ Daffy^{Duck},\ Yogi^{Bear}\}.$$

This C is again yin-functional, but it is not yang-functional. And of course, if we consider its transpose, $\overset{\leftrightarrow}{C}$, it is yang-functional but not yin-functional.

The composition of relations

Remember how, when we raised $\mathbb{N}$ up the tower to $\mathbb{Z}$, we gained the unary operator $-$? Well, when we raise the composition operator up from Couplets($\mathcal{G}$) to Relations($\mathcal{G}$), we fix the uncomfortable fact that compositions of couplets can be undefined. Neat, huh?

For couplets, we defined composition to be:

$$a^b \circ d^c = a^c \text{ if, and only if, } b = d$$

and when b was not equal to d, the outcome was undefined.

You may have wondered why we didn't just declare the outcome to be $\emptyset$. We would have liked to, but we couldn't bring ourselves to do it, because $\emptyset$ is not an element of the Cartesian product $\mathcal{G} \times \mathcal{G}$. We would have violated our own rules, and the mathematical umpire would have penalized us and deducted any points we thought we might have scored.

With the composition of relations, we can have $\emptyset$ be the result of the composition of two singleton sets of couplets, because relations are sets. Indeed we can define the outcome to be in the pattern of *the-flarn-of-the-clarp-is-the-clarp-of-the-flarns* or using real terminology, the composition of the relations is the relation of the (defined) compositions of couplets.

So consider this example:

$$\{a^b, b^c, c^d\} \circ \{b^c, d^e\} = \{a^c, c^e\}.$$

Mathematically, the only compositions of couplets that are defined are $a^b \circ b^c = a^c$ and $c^d \circ d^e = c^e$, so the result is $= \{a^c, c^e\}$.

And here's an example with real values, where we compose poets and painters, rather than poetry and paintings:

$$\{Blake^{Poet}, Dali^{Painter}, Byron^{Poet}\} \circ \{Painter^{Surrealist}, Poet^{Romantic}\}$$
$$= \{Blake^{Romantic}, Dali^{Surrealist}, Byron^{Romantic}\}.$$

To make it crystal clear, if we try the following composition,

$$Blake^{Poet} \circ Byron^{Poet}.$$

the result is undefined. However,

$$\{Blake^{Poet}\} \circ \{Byron^{Poet}\} = \emptyset$$

is slightly different, giving the empty set as the result. The point is that ascending the algebraic tower smooths the path nicely.

And, just in case you wondered about it, the composition of relations turns out to be associative, so for A, B, $C \in$ Relations($\mathcal{G}$):

$$A \circ (B \circ C) = (A \circ B) \circ C.$$

You might have expected this, because composition (when there were defined results) was associative in Couplets($\mathcal{G}$).

However, composition is not commutative – which is a damn shame, but what can you do? You were probably able to quickly work that out, assuming you remembered what "commutative" meant. If you didn't, we'll remind you: it means the result is the same whatever the order of the operands. So + is commutative because $a + b = b + a$, and $\circ$ is not because:

$$\{a^b, b^c\} \circ \{c^a\} = \{b^a\}, \text{ whereas } \{c^a\} \circ \{a^b, b^c\} = \{c^b\}$$

or alternatively:

$$\{Byron^{Poet}, Poet^{Mad\ man}\} \circ \{Mad\ man^{Byron}\} = \{Poet^{Byron}\},$$

whereas:

$$\{Mad\ man^{Byron}\} \circ \{Byron^{Poet}, Poet^{Mad\ man}\} = \{Mad\ man^{Poet}\}.$$

Practical application

We have described relations mathematically, so let's now discuss how they manifest themselves in the digital world of computers. Algebraically, many of the files held on a PC are relations.

For example, a photograph is usually stored as an image, maybe with some contextual data provided by the digital camera, such as when the photo was taken, where it was taken and what the camera settings were. In other words, it is a set of couplets. Digital photography standards, such as TIFF, define such relations explicitly.

Simple text files (.txt files) on a PC are relations. On the PC, they have the characteristics extension ".txt," and each line of text in the file, except the last, is terminated with the ASCII CR (carriage return) and LF (line feed). If we had a file called Example.txt with three lines of text, then the relation could be defined as:

$$E = \{txt^{filetype}, line1^{text}, line2^{text}, line3^{text}\}.$$

If we wished to treat the file algebraically, we would read it and transform it into a relation while we did so. In general, audio files and video files are handled in a similar way to .txt files, breaking up the data into chunks. And they, too, are relations, as are email files.

For all files that are commonly shared between different programs (text, images, graphics, video, music and so on) there are standard data formats invented by software companies or agreed upon by standards bodies. There have been efforts to formulate more general standards for this purpose, such as XML.

XML (it stands for the eXtensible Markup Language) is an open standard that provides a set of rules for encoding information (often documents) using tags. It is an open standard intended to be understood by humans and computers alike. Below is a very simple example:

```
<?xml version="1.0" encoding="UTF-8"?>
<note>
  <to> Reader</to>
  <from>Authors</from>
  <heading>Promotional Message</heading>
  <body>Data algebra is way cool!</body>
</note>.
```

With XML, it is standard for the first line to declare the XML version number and the encoding used. The rest of the information is data, delimited by tags, which can easily be translated into a relation called **Note**, as follows:

$$Note = \{Reader^{to}, Authors^{from}, Promotional\ Message^{heading},$$
$$Data\ algebra\ is\ way\ cool!^{body}\}.$$

If we wanted to select just the "to" and "body" from the relation we can simply construct another relation called **Select** as follows:

$$Select = \{to^{to}, body^{body}\}.$$

and then we could evaluate:

$$Note \circ Select = \{Reader^{to}, Data\ algebra\ is\ way\ cool!^{body}\}.$$

Let us now consider another common computer file, a CSV file. For this example, let's return to our set of superheroes A and their "powers" B. Each consists of just three elements, so that:

$$A = \{Thor,\ Hulk,\ Iron\ Man\}$$

and:

$$B = \{hammer,\ anger,\ suit\}.$$

We can take the Cartesian product $A \times B$ of these two sets. This would give us the set of 9 couplets displayed in the following array:

$A \times B$		B		
		hammer	*anger*	*suit*
A	*Thor*	$Thor^{hammer}$	$Thor^{anger}$	$Thor^{suit}$
	Hulk	$Hulk^{hammer}$	$Hulk^{anger}$	$Hulk^{suit}$
	Iron Man	$Iron\ Man^{hammer}$	$Iron\ Man^{anger}$	$Iron\ Man^{suit}$

That's fine, but it gives us six couplets we are not really interested in. The only ones we are interested in are those where our superheroes are married with the source of their powers. So imagine that we were provided these two sets, A and B, in the form of a CSV file on a computer, as follows:

Thor, Hulk, Iron Man
"hammer", "anger", "suit".

We are familiar with what a CSV file is. So we know that there is an implicit order in the elements of the two records in such a file. When parsing the records, as we read them from disk, we might turn them into ordered 3-tuples as follows:

$$C = \{1^{Thor},\ 2^{Hulk},\ 3^{Iron\ Man}\}$$
$$D = \{1^{hammer},\ 2^{anger},\ 3^{suit}\}.$$

That would allow us to transpose C to become $\overleftrightarrow{C}$ and compose it with D to give $\overleftrightarrow{C} \circ D$, which would look like this:

$\overleftrightarrow{C} \circ D$		D		
		1^{hammer}	2^{anger}	3^{suit}
$\overleftrightarrow{C}$	$Thor^1$	$Thor^{hammer}$		
	$Hulk^2$		$Hulk^{anger}$	
	$Iron\ Man^3$			$Iron\ Man^{suit}$

and now we have what we were hoping for:

Thor is united with his hammer,
Hulk is married to his characteristic ill temper,
and Iron Man is wearing his shiny yellow and red suit.

So, by composing two relations of cardinality 3, we end up with a relation of cardinality 3, that looks like this:

$$\{Thor^{hammer},\ Hulk^{anger},\ Iron\ Man^{suit}\}.$$

Diagonals

We need to clue you in about diagonals so you don't react adversely or even run off screaming when we tell you what the signature of Relations($\mathcal{G}$) is. And, to be honest, there is no need to run off screaming, because diagonals are fairly simple. Consider:

$\mathcal{G} \times \mathcal{G}$		$\mathcal{G}$			
		a	b	...	z
$\mathcal{G}$	a	a^a	a^b	...	a^z
	b	b^a	b^b	...	b^z
	...	...	...	...	...
	z	z^a	z^b	...	z^z

This Cartesian product is the great and glorious couplet factory that manufactures every possible couplet of data that will ever find its way into a computer.

Now consider $\mathfrak{P}(\mathcal{G} \times \mathcal{G})$. It is a huge mountain of sets containing all those couplets as sets, in every collection of couplets you can imagine. And every collection of couplets in $\mathfrak{P}(\mathcal{G} \times \mathcal{G})$ is a relation, which means that the set of couplets which make up the main diagonal of the above display is also a relation.

We denote this relation as D_G, defining it as follows:

$$D_G := \{a^a \in \, G \times G : a \in G\},$$

which means that:

$$D_G = \{a^a, b^b, \ldots, z^z\}.$$

Here we have to pedantically make the point that the table shown above is a convenient visual artifice that illustrates the elements of the diagonal in a convenient way for the reader. The importance of D_G is that it serves as the identity for the composition in Relations(G):

$$R \circ D_G = R = D_G \circ R.$$

It turns out that subsets of D_G are also very important in Relations(G). We hope you'll realize later when we start to manipulate data more aggressively than we have done so far.

We note, in passing, that:

$$yin \, (D_G) = yang \, (D_G) = G.$$

The signature of Relations(G)

The signature of Relations(G) is:

$$[\mathfrak{P}(G \times G), \{[\circ, D_G]\}, \{\leftrightarrow\}].$$

This announces that all the relations deriving from the genesis set G are elements of the power set $\mathfrak{P}(G \times G)$ and that Relations(G) includes a binary operator $\circ$, whose identity is D_G and a unary operator $\leftrightarrow$.

To be clear, the **genesis set** of Relations(G) is the same G that gave birth to Couplets(G), but its **ground set** is the glorious power set $\mathfrak{P}(G \times G)$, which has implications that you might not have yet realized. So to remind you about power sets...

If we have a set $A = \{a, b, c\}$, the power set $\mathfrak{P}(A)$ is:

$$\{\emptyset, \{a\}, \{b\}, \{c\}, \{a, b\}, \{a, c\}, \{b, c\}, \{a, b, c\}\}.$$

This consists of every possible subset of A, including the set A itself and the empty set $\emptyset$. When we construct the power set of the Cartesian product $\mathcal{G} \times \mathcal{G}$, $\mathfrak{P}(\mathcal{G} \times \mathcal{G})$, we get every possible set of couplets $\mathcal{G}$ can furnish. The elements of $\mathcal{G} \times \mathcal{G}$ are couplets, but the elements of $\mathfrak{P}(\mathcal{G} \times \mathcal{G})$ are all sets of couplets. So they are relations.

What the signature of Relations($\mathcal{G}$) fails to mention is that $\mathfrak{P}(\mathcal{G} \times \mathcal{G})$, by its very nature, supports a set-based Boolean algebra. So, in climbing the tower, we have not only taken two useful operators with us, we have purloined the set operators, union $\cup$, intersection $\cap$, complement $'$ and the subset relation $\subset$. They are all valid for Relations($\mathcal{G}$), but by mathematical convention, rather than through modesty, we do not draw attention to that in the signature.

A word about functions

If you hark back to Chapter 4, you may remember we warned you about functions. Very specifically, we warned you that all those lovely well-behaved $y = f(x)$ types of functions that you enjoyed in school, or possibly at university, are not a feature of data algebra. You are probably interested to know why. The answer is not complicated, but it is important.

Those $y = f(x)$ functions are just one example of a function. Mathematically, a function is something that, when provided with an input, gives you a unique output. So if we have $f(x) = 3x + 2$, and we provide the input $x = 5$, then the output will be 17. And because we can take values of x very, very close to each other, when we plot the output on paper with Cartesian coordinates, it will look like a straight line. And if the function was $f(x) = x^2$, the graph would look like a parabola. Numeric algebra, with its beautiful polynomial functions, can easily seduce you into believing that all mathematical life can be like that. It cannot.

We can also have a function f(person) = SSN, where SSN is a person's social security number. This doesn't conveniently translate into a graph with axes. Neither is there some known calculation or procedure can be applied in some fashion, where if you plug a person into the function it naturally calculates the SSN. All you have is a set of inputs (people) and a set of outputs (SSNs). In fact, what you have if you take the input and the output together is a set of couplets, each couplet of which is of the form: $person^{SSN}$.

So when you encounter other data algebra functions later in this book, please remember that.

Ordered n-tuples

We shall round up this chapter by considering "the mysterious case of the missing Cartesian product." We noted earlier – you may have missed it, but believe me, we did draw attention to it – that Cartesian products of the form $A \times B \times C$ are not kosher. In fact, to be precise, we wrote "$A \times B \times C$ is notational nonsense unless $A = \emptyset$ or $B = \emptyset$ or $C = \emptyset$."

That's all very well, but what if we want to create some bona fide 3-tuples? Let's examine a set of ordered 3-tuples that we can construct from the set U. We start with an ordered set of cardinality 3, the set of integers $\{1, 2, 3\}$. An ordered 3-tuple is an element of the Cartesian product:

$$\mathfrak{P}(\{1, 2, 3\} \times U).$$

A power set of a Cartesian product is not the simplest set in the world. It is peppered with many couplets that we have very little interest in here. Nevertheless, it will contain some sets of the form $\{1^a, 2^b, 3^c\}$ where $a, b, c \in U$. That is, it will contain some sets of three couplets that are ordered 3-tuples.

Now let $A \cup B \cup C \subset U$. We can then designate $A \times B \times C$ to be:

$$\{\{1^a, 2^b, 3^c\} \in \mathfrak{P}(\{1, 2, 3\} \times U): a \in A, b \in B \text{ and } c \in C\}.$$

This resolves the mysterious case of the missing Cartesian product, $A \times B \times C$, although to be honest, you probably had no idea that there was a mysterious case in progress, even though there was. But nevermind, we have a capability here to create ordered pairs, ordered triples and ordered n-tuples – all of which are sets of couplets. The general case is:

If $A \cup B \cup C \cup ... \cup N \subset U$, we can designate $A \times B \times C \times... \times N$ to be

$$\{\{1^a, 2^b, 3^c..., n^z\} \in \mathfrak{P}(\{1, 2, 3 ..., n\} \times U): a \in A, b \in B, c \in C, ..., z \in N\}.$$

And, please note, those $1^a, 2^b, 3^c...$ etc. are elements of a set that we happen to have written in a specific order but could be written in any order. It is the yin set of the n-tuple, $\{1, 2, 3 ..., n\}$, that imposes an order, and it does so because it is an initial segment of the natural numbers $\mathbb{N}$, not because of any apparent arrangement of the list.

If you have a sharp mathematical eye, you may have noticed that we actually introduced ordered n-tuples on page 64, without deigning to explain them, when we chose to enforce an order on the following collection of text:

$$\{\{In^{preposition}\}^1, \{the^{definite\text{-}article}\}^2, jungle^{noun}\}^3, \{,^{comma}\}^4, \{the^{definite\text{-}article}\}^5,$$
$$\{mighty^{adjective}\}^6, \{jungle^{noun}\}^7, \{,^{comma}\}^8, \{the^{definite\text{-}article}\}^9, \{lion^{noun}\}^{10},$$
$$\{sleeps^{verb}\}^{11}, \{tonight^{adverb}\}^{12}\}^{lyrics}.$$

Now you have an explanation. We can, with very little effort, enable the formation of ordered n-tuples by including a subset of the natural numbers in any genesis set we choose to create. And, in the sphere of software, there is good reason to do that, because software is often required to sort data into a particular order.

More importantly, we have arrived at the point where we can confidently declare what an ordered pair is. It's an ordered 2-tuple – a certain set of two couplets – nothing more, nothing less.

Taking Stock

This chapter provided additional information about couplets, and then climbed the tower into the algebra of relations. The following bullet points summarize its content:

- We can use the natural numbers within a Cartesian product to create ordered sets, ordered n-tuples.

- We define the functions *yin* and *yang* as follows: *yin* $(a^b) = a$ and *yang* $(a^b) = b$. These functions also can be applied to Relations($\mathcal{G}$).

- We defined the **yin-set** and the **yang-set** as follows: If we apply the *yin* or *yang* function to a relation *R*, we call the result the yin-set of *R*, or the yang-set of *R*, accordingly.

- The transpose of a relation *R* is as follows:

$$\overleftrightarrow{R} = \{\, b^a \in B \times A : \overleftrightarrow{b^a} \in A \times B \}.$$

- We defined **yin-functional**: If $a^b \in R$ and $a^c \in R$ imply that $b = c$ then *R* is said to be **yin-functional.**

- We defined **yang-functional**: *R* is **yang-functional** if $\overleftrightarrow{R}$ is yin-functional.

- Composition for Couplets($\mathcal{G}$): $a^b \circ d^c = a^c$ if, and only if, $b = d$.

- Composition is associative but not commutative.

- The diagonal of $\mathcal{G}$ is defined as $D_{\mathcal{G}} = \{a^a \in \mathfrak{P}(\mathcal{G} \times \mathcal{G}), a \in \mathcal{G}\}$. It is the identity for composition.

- The signature of Relations($\mathcal{G}$) is $[\mathfrak{P}(\mathcal{G} \times \mathcal{G}), \{[\circ, D_{\mathcal{G}}]\}, \{\leftrightarrow\}]$. The set operators, union $\cup$, intersection $\cap$, complement $'$ and the subset relation $\subset$ are all valid for Relations($\mathcal{G}$).

- We define n-tuples as follows: If $A \cup B \cup C \cup ... \cup N \subset U$, we designate $A \times B \times C \times ... \times N$ to be

$$\{\{1^a, 2^b, 3^c ..., n^z\} \in \mathfrak{P}(\{1, 2, 3 ..., n\} \times U): a \in A, b \in B, c \in C, ..., z \in N\}.$$

Chapter 7: The Gathering of the Clans

There are nine and sixty ways of constructing tribal lays, And every—single—one—of—them—is—right!

~ Rudyard Kipling

———⸳∞⸳———

IN THE EARLY DAYS of the relational model of data, many of its enthusiastic advocates insisted that there was a "right structure" for data. This "right structure" was called "third normal form." The "normalizing" of data to "third normal form" was a fairly simple data modeling process that aimed to minimize redundancy (i.e., data duplication) and incorrect data dependencies, so that update anomalies could not occur.

This data modeling approach was gradually eclipsed by Peter Chen's more natural Entity Relationship (ER) model, which produced a very similar outcome and gradually became dominant. Such data models were the work horses for the OLTP systems they helped to build through the 1980s and 1990s. However, by the late 1990s, as large companies began constructing data models for the large collections of data that inhabited a data warehouse, a fully normalized data structure proved impractical. This was due to the multitude of awkwardly slow database JOIN operations it engendered.

Consequently, an alternative approach was adopted. The data to be held in the data warehouse first needed to be normalized to third normal (or some variant thereof), and then the normalized model needed to be compressed into far fewer tables to create what was called a "snowflake schema" or even a "star schema." If this process of normalizing in order to denormalize has you reciting *The Grand Old Duke of York*, you are not alone.

Certainly there are data structures that are easier for human beings to understand. Technically, such models of data are usually referred to as logical models. These models hide many of the details of data that a computer needs to know. Sadly, humans have never achieved complete agreement on exactly what the logical model should be for a given collection of information. There are debates and disagreements.

There is also what is called a physical model of data, which arranges the data in a form that a computer can know and love. In truth, this is not so much a model as a data structure that is implemented for a given collection of information. However, it is often the case that this structure derives from a software engineer's model of the way a computer can most effectively process the data.

Here there are pragmatic reasons why there may be disagreement on the best data structure. Once data enters the computer, there is no knowing for certain how it will be used. All a software engineer can do is estimate how it may be used and structure it accordingly. The optimal data structure will be the one that consumes least computer resources while meeting the response time or throughput goals of the application. Because computer hardware is continually evolving – the power of CPUs continually improves, memory speeds up, SSD speed up, networks speed up and so on – the optimal data structure for any application itself changes with time.

In practice, the data structures that computers know and love are distinctly different from the data structures that the humans know and love (or disagree about). Consequently, in building any application, it is necessary to map the logical model to the optimal physical data structure so that computers and human beings can live together in harmony.

From a data algebraist's perspective, there is no correct data structure. Certainly there can be incorrect structures that are algebraically deficient, but that deficiency will find its origin in human error. Where that is not the case, in situations where competing logical data models are fighting it out, we are happy to proclaim that every-single-one-of-them-is-right.

As for physical data structures, the goal is optimization, and mathematicians have skin in that game – or better put, optimization is the soul of mathematics. There is very good reason to believe that optimizing physical digital data structures, no matter what their context, is a work that data algebra will dominate in the long run.

In the meantime, almost all the world's data is structured in some way or other, and, no matter how it is structured, its structure can be captured in an algebraic form, as we are gradually demonstrating in this book. Some of the data, perhaps even most of it, may be algebraically deficient, but that can be fixed – and hopefully in time it will be. The rest of it is in the right form for the application of data algebra, but it may not be in the optimal form.

Relations beget clans

A clan is a set of relations. That's how it works in Scotland and that's how it works in data algebra. If the relations are similar as they are in the example below, then the clan is similar to a table in a relational database, and a clan can certainly express such a table.

This is a clan $\mathbb{O}$, of recent Oscar-winning actresses:

$$\{\{2011^{year}, Meryl\ Streep^{actress}, The\ Iron\ Lady^{movie}\},$$
$$\{2012^{year}, Jennifer\ Lawrence^{actress}, Silver\ Linings\ Playbook^{movie}\},$$
$$\{2013^{year}, Cate\ Blanchett^{actress}, Blue\ Jasmine^{movie}\},$$
$$\{2014^{year}, Julianne\ Moore^{actress}, Still\ Alice^{movie}\}\}.$$

You may prefer to view it using the visual artifice of a table:

Table 1: Table O of Oscar-winning actresses		
year	**actress**	**movie**
2011	Meryl Streep	The Iron Lady
2012	Jennifer Lawrence	Silver Linings Playbook
2013	Cate Blanchett	Blue Jasmine
2014	Julianne Moore	Still Alice

If you prefer that, then for algebraic accuracy you need to remember that the last four rows of the table are just the **yin** sets of the relations they represent. The column headings are, believe it or not, **yang** ($\mathbb{O}$). This provides us with an excellent excuse for explaining what the **yin** and **yang** of a clan are.

The yin-set of a clan $\mathbb{C}$, **yin**($\mathbb{C}$), is the union of the yin sets of the relations in $\mathbb{C}$. We can write this:

$$yin(\mathbb{C}) = \bigcup_{R \in \mathbb{C}} yin(R).$$

In our example, **yin**($\mathbb{O}$) is:

{2011, Meryl Streep, The Iron Lady, 2012, Jennifer Lawrence, Silver Linings Playbook, 2013, Cate Blanchett, Blue Jasmine, 2014, Julianne Moore, Still Alice}.

You already know what **yang**($\mathbb{O}$) is. The general algebraic expression is:

$$yang(\mathbb{C}) = \bigcup_{R \in \mathbb{C}} yang(R).$$

Clans do not have to be conveniently structured so that they fit neatly into a table. We can, for example, create a clan by simply lumping together two of our previous examples of relations, as follows:

$$\mathbb{A} = \{\{Thor^{hammer}, Hulk^{anger}, Iron\ Man^{suit}\},$$
$$\{Bugs^{Bunny}, Daffy^{Duck}, Wiley^{Coyote}\}\}.$$

It is hard to imagine a good reason to do that, but nevertheless if you do it, you have a set of relations, and that's a clan; you cannot shoehorn it into a table with a heading and rows, unless you are willing to accept a table containing empty cells like this:

Table 2: Table with empty cells					
hammer	**anger**	**suit**	**Bunny**	**Duck**	**Coyote**
Thor	Hulk	Iron Man			
			Bugs	Daffy	Wiley

We can also build a clan directly from a relation. For example, if we take the simple relation:

$$B = \{Bugs^{Bunny}, Daffy^{Duck}, Wiley^{Coyote}\}$$

and partition it into two subsets, and then collect them together in a set as shown:

$$\mathbb{B} = \{\{Bugs^{Bunny}, Daffy^{Duck}\},$$
$$\{Wiley^{Coyote}\}\}$$

we have created a clan from the relation. Similarly, if we take the union of those two relations, the clan is immediately demoted to being a relation. That's the joy of the algebraic tower. In some circumstances you can climb up the tower and then climb back down, as long as you don't break any of the rules.

Transposition and composition

You will be comforted to know that operations of transposition and composition lift from Relations($\mathcal{G}$) to Clans($\mathcal{G}$) with the minimum of fuss. Let's show and tell:

If $\mathbb{A} = \{\{Thor^{hammer}, Hulk^{anger}, Iron\ Man^{suit}\}, \{Bugs^{Bunny}, Daffy^{Duck}, Wiley^{Coyote}\}\}$, then

$$\overleftrightarrow{\mathbb{A}} = \{\{hammer^{Thor}, anger^{Hulk}, suit^{Iron\ Man}\},$$
$$\{Bunny^{Bugs}, Duck^{Daffy}, Coyote^{Wiley}\}\}.$$

Now that we've demonstrated it, we can confidently chant that the transpose of the clan is the clan of the transpositions, or *the-flarn-of-the-clarp-is-the -clarp-of-the-flarns*.

The same applies with composition: the composition of two clans is the clan of the compositions.

Consider a very-simple-for-the-sake-of-illustration clan $\mathbb{B}$:

$$\mathbb{B} = \{\{anger^{management}, Duck^{Dynasty}\}, \{suit^{yourself}\}\}.$$

$$\mathbb{A} \circ \mathbb{B} = \{\{Thor^{hammer}, Hulk^{anger}, Iron\ Man^{suit}\}, \{Bugs^{Bunny}, Daffy^{Duck},$$
$$Wiley^{Coyote}\}\} \circ \{\{anger^{management}, Duck^{Dynasty}\}, \{suit^{yourself}\}\}$$

$$= \{\{Hulk^{management}\}, \{Iron\ Man^{yourself}\}, \{Daffy^{Dynasty}\}\}.$$

Particularly take note of this point: we can "compose" a couplet a^b or a relation R with a clan simply by promoting the couplet to be a clan by wrapping it in double braces as $\{\{a^b\}\}$ or by promoting the relation to be a clan by wrapping it in single braces as $\{R\}$. Mathematically, as shorthand, because we know we can do this, we tend not to show the braces.

Yin-functional and yang-functional

Just like transposition and composition, the concepts of yin-functional and yang-functional lift naturally to clans. We can define them as follows:

$\mathbb{C}$ *is yin-functional if each of its elements, i.e., each of its relations is yin-functional.*

Just to remind you of what yin-functional means, a relation R is yin-functional if $a^b \in R$ and $a^c \in R$ imply that $b = c$.

$\mathbb{C}$ *is yang-functional if $\overleftrightarrow{\mathbb{C}}$ is yin-functional.*

The clans $\mathbb{A}$ and $\mathbb{B}$ above are yin-functional since in both cases, their two relations are yin-functional. The clan $\mathbb{D}$ below is an example of a clan that is neither yang-functional nor yin-functional:

$$\mathbb{D} = \{\{hammer^{Thor}, blonde^{Thor}, suit^{Iron\ Man}\},$$
$$\{Bunny^{Bugs}, Duck^{Daffy}, Duck^{soup}\}\}.$$

The "first" relation is not yin-functional because of the *Thor* couplets, and the "second" relation is not yang-functional because of the *Duck* couplets. We are obliged to wrap first and second in quotes, tedious though it may seem, because $\mathbb{D}$ is a set, and hence there is no specific order to those relations.

We are now going to introduce two more yin/yang terms as a special reward for your perseverance in getting this far through the book. You are thinking that you need two more such yin/yang terms like a hole in the head, so let us try to convince you otherwise. The new terms, by the way, are **yang-regular** and **yin-regular**.

The clan $\mathbb{C}$ is yang-regular if for each $\mathbf{R} \in \mathbb{C}$, $\mathbf{R}$ is yang-functional and yang($\mathbf{R}$) = yang($\mathbb{C}$).

Our previous clan, $\mathbb{O}$, of recent Oscar-winning actresses, is yang-regular.

$\{\{2011^{year}, \text{Meryl Streep}^{actress}, \text{The Iron Lady}^{movie}\},$
$\{2012^{year}, \text{Jennifer Lawrence}^{actress}, \text{Silver Linings Playbook}^{movie}\},$
$\{2013^{year}, \text{Cate Blanchett}^{actress}, \text{Blue Jasmine}^{movie}\},$
$\{2014^{year}, \text{Julianne Moore}^{actress}, \text{Still Alice}^{movie}\}\}.$

This mean that the clan can be represented as a nice regular table with the headings: *year, actress and movie*. Take any one of the four relations in $\mathbb{O}$ and its yang-set is: {*year, actress, movie*}. Each of these relations is yang-functional, and if you take the union of their yang sets you get *yang*($\mathbb{O}$), i.e:

$$\textit{yang}(\mathbb{O}) = \bigcup_{\mathbf{R} \in \mathbb{O}} \textit{yang}(\mathbf{R}).$$

It is clear that for every $\mathbf{R}$, $\textit{yang}(\mathbf{R}) = \textit{yang}(\mathbb{O}) = \{year, actress, movie\}$.

As you may have guessed:

$\mathbb{C}$ is yin-regular if $\overset{\leftrightarrow}{\mathbb{C}}$ is yang-regular.

Take a look at $\overset{\leftrightarrow}{\mathbb{O}}$ below:

$\{\{year^{2011}, actress^{\text{Meryl Streep}}, movie^{\text{The Iron Lady}}\},$
$\{year^{2012}, actress^{\text{Jennifer Lawrence}}, movie^{\text{Silver Linings Playbook}}\},$
$\{year^{2013}, actress^{\text{Cate Blanchett}}, movie^{\text{Blue Jasmine}}\},$
$\{year^{2014}, actress^{\text{Julianne Moore}}, movie^{\text{Still Alice}}\}\}.$

The yang-set of each of the relations in $\overset{\leftrightarrow}{\mathbb{O}}$ is different and hence it is not the case that for every $\mathbf{R}$, $\textit{yang}(\mathbf{R}) = \textit{yang}(\overset{\leftrightarrow}{\mathbb{O}})$.

The Tier of Clans($\mathcal{G}$)

No doubt you are itching to know about the ground set of the clan tier of the tower. Aside from calamine lotion, the only known cure for that itch is to tell you about it.

First of all, we can promise you that it has a lot more in it than the comparatively anorexic $\mathfrak{P}(\mathcal{G} \times \mathcal{G})$. The ground set of clans is $\mathfrak{P}(\mathfrak{P}(\mathcal{G} \times \mathcal{G}))$, or as we prefer to write it, $\mathfrak{P}^2(\mathcal{G} \times \mathcal{G})$.

To get an idea of how large a power set of a power set is, consider the almost unpopulated set $\{a, b\}$. If we write that out, it looks like this:

$$\mathfrak{P}(\{a, b\}) = \{\emptyset, \{a\}, \{b\}, \{a, b\}\}.$$

Writing out the power set of a power set is the kind of thing you are made to do for punishment if you behave badly in mathematics class and have to stay behind after school. You can see how difficult that can be just from glancing at the power set of a power set that has just two elements. It looks like this:

$\mathfrak{P}^2(\{a, b\}) = \{\emptyset,$

$\{\emptyset\}, \{\{a\}\}, \{\{b\}\}, \{\{a, b\}\},$

$\{\emptyset, \{a\}\}, \{\emptyset, \{b\}\}, \{\emptyset, \{a, b\}\}, \{\{a\}, \{b\}\}, \{\{a\}, \{a, b\}\}, \{\{b\}, \{a, b\}\},$

$\{\emptyset, \{a\}, \{b\}\}, \{\emptyset, \{a\}, \{a, b\}\}, \{\emptyset, \{b\}, \{a, b\}\}, \{\{a\}, \{b\}, \{a, b\}\},$

$\{\emptyset, \{a\}, \{b\}, \{a, b\}\}\}.$

The 16 sets above are listed in order of increasing cardinality: first the empty set, then sets with one element (there are four), then sets with two elements (there are six), then sets with three elements (there are four) and finally the set with four elements. The actual number of elements in this set is given by:

$$|\mathfrak{P}^2(\{a, b\})| = 2^{(2^2)} = 16.$$

Having to write out the power set of the power set of a set with three elements qualifies as a cruel and unusual punishment under international law. You really don't have to have much of a genesis set before the cardinality of the power set of the power set gets astronomic. It doesn't just increase exponentially, it increases exponentially, exponentially. You need to take the logarithm of the logarithm of the cardinality to keep it under control.

And now for something completely different:

When we climb from the algebra of relations to the algebra of clans, we bump up against a symbol problem. You may have noticed that when we ascended from couplets to relations and then from relations to clans, we took care to define what we meant by transposition ($\leftrightarrow$) and composition ($\circ$) in each new algebra. Because the effect of each operator was similar from one algebra to another, there was no confusion created by using exactly the same symbol.

When we climbed up into the algebra of relations we happily started to use the union, $\cup$, and intersection, $\cap$, operators not just because they applied, given that Relations($\mathcal{G}$) was a set algebra, but also because we were not using those symbols for any other purpose.

Well, when we climb up into Clans($\mathcal{G}$) the union, $\cup$, and intersection, $\cap$, operators pose a minor problem for us. As we remarked way back in Chapter 3, when you climb an algebraic tower, sometimes you don't get the outcome you hoped for.

So for the moment let us consider the relatively simple power set of a power set: $\mathfrak{P}^2(\{a, b, c\})$, and the sets A = $\{\{a, b\}, \{a, c\}\}$ and B = $\{\{a\}, \{a, b, c\}, \{a, c\}\}$ that are elements of $\mathfrak{P}^2(\{a, b, c\})$. The union of these two sets is:

$$A \cup B = \{\{a, b\}, \{a, c\}\} \cup \{\{a\}, \{a, b, c\}, \{a, c\}\}$$

$$= \{\{a, b\}, \{a, c\}, \{a\}, \{a, b, c\}\}$$

and the intersection is:

$$A \cap B = \{\{a, b\}, \{a, c\}\} \cap \{\{a\}, \{a, b, c\}, \{a, c\}\}$$

$$= \{\{a, c\}\}.$$

You may not be aware of it yet, but there is another kind of union, called a cross-union, and another kind of intersection, called a cross-intersection. We bet you just cannot wait to find out what they are. The cross-union can be defined like this:

> *The cross-union of two sets of sets is the set of all possible unions of sets from the original sets.*

That's an awful thing to have to read, but we wrote it down anyway. If we had been just focusing on clans, it would have made much easier reading. We could have written a much simpler sentence like this:

The cross-union of two clans is the set of all possible unions of relations from within the respective clans.

The official definition we wrote is blessed with generality but cursed by inscrutability. Nevertheless, it's possible to get the idea if we follow the words and follow an example. We use the symbol ▼ to indicate the cross union. So, remembering that in this instance, A = {{a,b}, {a, c}} and B = {{a}, {a, b, c}, {a, c}} let's evaluate:

$$A ▼ B = \{\{a, b\} \cup \{a\}, \{a, b\} \cup \{a, b, c\}, \{a, b\} \cup \{a, c\}, \{a, c\} \cup \{a\},$$
$$\{a, c\} \cup \{a, b, c\}, \{a, c\} \cup \{a, c\}\}.$$

Take a look. That really is "the set of all possible unions of sets from the original sets." And if you work it out, it gives the result:

$$A ▼ B = \{\{a, b\}, \{a, c\}, \{a, b, c\}\}.$$

If we compare the two results, clearly $A \cup B \neq A ▼ B$ in this particular case. It can be the same in some instances, but it is seldom so.

Naturally, the cross-intersection, which we denote by the symbol, ▲, arrives with an equally inscrutable definition:

The cross intersection of two sets of sets is the set of all possible intersections of sets from the original sets.

So, let's evaluate:

$$A ▲ B = \{\{a, b\} \cap \{a\}, \{a, b\} \cap \{a, b, c\}, \{a, b\} \cap \{a, c\}, \{a, c\} \cap \{a\},$$
$$\{a, c\} \cap \{a, b, c\}, \{a, c\} \cap \{a, c\}\}.$$

Again, that really is "the set of all possible intersections of sets from the original sets." Work it out and you get:

$$A ▲ B = \{\{a\}, \{a, b\}, \{a, c\}\}.$$

Again, if we compare the two results, clearly $A \cap B \neq A ▲ B$. And again, it can be the same in some instances, but it is seldom so.

We can now apply all of this to couplets, relations and clans. So let's have some similar examples to work with, clans that are elements of $\mathfrak{P}^2(\{a^x, b^y, c^z\})$. Let $\mathbb{A} = \{\{a^x, b^y\}, \{a^x, c^z\}\}$ and $\mathbb{B} = \{\{a^x\}, \{a^x, b^y, c^z\}, \{a^x, c^z\}\}$ and let's calculate the garden variety union:

$$\mathbb{A} \cup \mathbb{B} = \{\{a^x, b^y\}, \{a^x, c^z\}\} \cup \{\{a^x\}, \{a^x, b^y, c^z\}, \{a^x, c^z\}\}$$

$$= \{\{a^x, b^y\}, \{a^x, c^z\}, \{a^x\}, \{a^x, b^y, c^z\}\}.$$

And now the cross-union:

$$A \blacktriangledown B = \{\{a^x, b^y\} \cup \{a^x\}, \{a^x, b^y\} \cup \{a^x, b^y, c^z\}, \{a^x, b^y\} \cup \{a^x, c^z\},$$
$$\{a^x, c^z\} \cup \{a^x\}, \{a^x, c^z\} \cup \{a^x, b^y, c^z\}, \{a^x, c^z\} \cup \{a^x, c^z\}\}.$$

In the above equation we really are taking the set of all possible unions of relations from within the respective clans. This gives us:

$$A \blacktriangledown B = \{\{a^x, b^y\}, \{a^x, c^z\}, \{a^x, b^y, c^z\}\}.$$

If we compare the two results, clearly $A \cup B \neq A \blacktriangledown B$. The union and the cross-union are definitely different beasts of burden, and both of them will serve us well in $\mathfrak{P}^2(\mathcal{G} \times \mathcal{G})$.

Here's the intersection:

$$A \cap B = \{\{a^x, b^y\}, \{a^x, c^z\}\} \cap \{\{a^x\}, \{a^x, b^y, c^z\}, \{a^x, c^z\}\}$$

$$= \{\{a^x, c^z\}\}.$$

And here's the cross-intersection:

$$A \blacktriangle B = \{\{a^x, b^y\} \cap \{a^x\}, \{a^x, b^y\} \cap \{a^x, b^y, c^z\}, \{a^x, b^y\} \cap \{a^x, c^z\},$$
$$\{a^x, c^z\} \cap \{a^x\}, \{a^x, c^z\} \cap \{a^x, b^y, c^z\}, \{a^x, c^z\} \cap \{a^x, c^z\}\}$$

$$= \{\{a^x\}, \{a^x, b^y\}, \{a^x, c^z\}\}.$$

And, $A \cap B \neq A \blacktriangle B$. Which you probably expected, recognizing that we deliberately chose A and B to give results that didn't match for union and intersection.

The view from the tower

In summary, we ascended the tower of data algebra and discovered to our annoyance that the state of the union (and the intersection) were not as we might have hoped. Down there in $\mathfrak{P}(\{a^x, b^y, c^z\})$, we could happily take the union of $\{a^x, b^y\}$ with $\{b^y, c^z\}$ to get $\{a^x, b^y, c^z\}$. Climb the giddy tower and, what do you know, suddenly $\{\{a^x, b^y\}\} \cup \{\{b^y, c^z\}\} = \{\{a^x, b^y\}, \{b^y, c^z\}\}$.

In Relations($\mathcal{G}$), when we evaluate $\{a^x, b^y\} \cup \{b^y, c^z\}$, we get $\{a^x, b^y, c^z\}$, but in that medieval corner of the universe we process one relation at a time. With clans, we want to unite or divide hundreds, maybe even thousands, of relations with one blow. And this is what our brave new operators $\blacktriangledown$ and $\blacktriangle$

will do for us. Let's illustrate a couple of simple possibilities by considering the clan $\mathbb{A}$, where:

$$\mathbb{A} = \begin{array}{l} \{\{a^1, a^2, a^3, a^4, a^5\}, \\ \{b^1, a^4, b^2\}, \\ \{c^1, c^2, a^5, c^3, c^6, a^6\}, \\ \{d^1, d^2\}, \\ \{e^1, d^1\}\}. \end{array}$$

If we want to insert a couplet, say, $abcde^1$, into each of the relations in this clan, all we need to do is construct a clan $\mathbb{B}$, where $\mathbb{B} = \{\{abcde^1\}\}$, and calculate:

$$\mathbb{A} \blacktriangledown \mathbb{B} = \begin{array}{l} \{\{abcde^1, a^1, a^2, a^3, a^4, a^5\}, \\ \{abcde^1, b^1, a^4, b^2\}, \\ \{abcde^1, c^1, c^2, a^5, c^3, c^6, a^6\}, \\ \{abcde^1, d^1, d^2\}, \\ \{abcde^1, e^1, d^1\}\}. \end{array}$$

Now consider the situation where we want to know whether any of the couplets a^5, c^4, d^2 or d^3 occur as an element of a relation in clan $\mathbb{A}$, then we can construct a clan $\mathbb{C} = \{\{a^5\}, \{c^4\}, \{d^2\}, \{d^3\}\}$ and calculate $\mathbb{A} \blacktriangle \mathbb{B}$. The result will be the clan $\{\{a^5\}, \{d^2\}\}$, which demonstrates that a^5 and d^2 can be found in relations in that clan.

The cross-intersection and cross-union turn out to be extremely useful operators. You will meet with them again.

The variety of data structures

In simple terms, we began with the algebra of couplets, which was quite limited, but nevertheless introduced us to the fundamental unit of data. We climbed up to the algebra of relations, which introduced an algebraic data structure that can represent a simple one-record file or a simple record in a database. Now that we have ascended to clans, we have attained an algebraic data structure that can represent a multi-record file or a table in a relational database, or a list, or an XML file with repeating structures.

We are now about to turn our attention to graphs, which are another data structure that can be represented by relations and clans, but before we do so, it is worth remarking that once we have demonstrated the algebraic representation of graphs, then we have discussed every common data structure in the field of computing (records, data sets, ordered sets, lists, arrays, associative arrays and graphs).

This does not mean that we have finished our ascent of the algebraic towers – we have not. Once we have had our say about clans, we will have a few comments about the collection of clans that are called hordes. But first, let's talk about graphs.

Data Algebra and Graphs

Mathematically, a graph is a structure that models pair-wise relationships between objects. The convention is to use the term "node" for the objects and the term "edge" for the connecting lines. Graphs can be *directed* in the sense that the edges indicate a direction, illustrated visually by an arrow head, or *simple,* where the edges do not indicate direction; they can also be *labeled* in the sense that there is data (a label) attached to each edge, or they can be unlabeled.

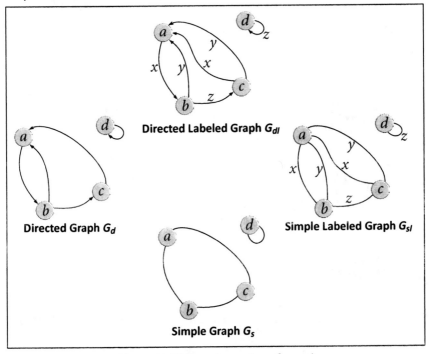

Figure 7: Different species of graph

So there are four possibilities:

1. **Directed labeled graphs**
2. **Directed graphs**
3. **Simple labeled graphs**
4. **Simple graphs**

Graphs can be represented visually, with edges connecting nodes of the graph, as illustrated in Figure 7. Such representations do not scale well, but they are useful for illustrative purposes.

There is a whole mathematics of graph theory which, if you have the time and the enthusiasm, you can investigate to your heart's content. Here we are interested primarily in graphs as data. We have a specific interest in this for two reasons:

1. Some data is better organized (for human consumption) as a graph or network of nodes.

2. Graphs expressed algebraically provide an effective way to represent semantic relationships in natural language.

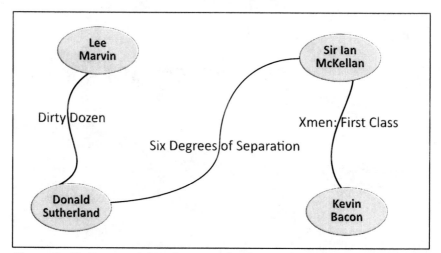

Figure 8: Six Degrees of Kevin Bacon (example of a simple labeled graph)

Those of you who are familiar with database technology may know that relational databases are not particularly suited to graphical applications. Consequently, there are graph database products that specialize in such applications. They focus on analyzing networks of relationships between people along the lines of the popular parlour game "Six Degrees of Kevin Bacon," which is based on the relationship "who was in a movie with whom?"

Our example in Figure 8 shows how Lee Marvin and Kevin Bacon are connected by various movies including a movie called (ironically) *Six Degrees of Separation*. Of course, this is simply a single answer among

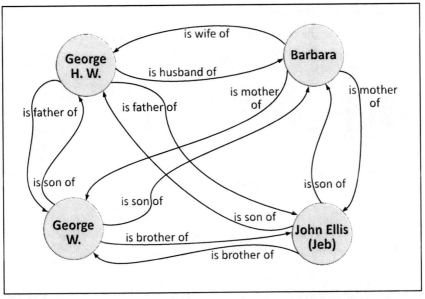

Figure 9: An example of a directed labeled graph

many possible answers drawn from a huge graph based on all the actors and actresses who ever appeared in a movie and the movies they appeared in.

The illustration in Figure 9 shows a much busier graph, with four nodes depicting the Bush family relationships between George H. W., Barbara, George W. and John Ellis (Jeb).

This graph already seems to have a few too many edges for comfort, but we could easily make it more complex by including more edges between the nodes, representing relations such as "who is older than whom," "who is taller than whom," "who is a better golfer than whom," and so on. Alternatively, we could include all of the Bush children who have never run for president.

In the first graph (Figure 8), the edges represent movies in which the two actors appeared, and the nodes represent the actors. In the second graph (Figure 9), the edges in the graph have arrowheads indicating the direction of the relationship expressed in the label and the nodes are the family members. The two data graphs are different and have different algebraic representations.

Directed labeled graphs

These graphs can have as many directed edges as you wish between nodes. We represent the adjacent directed labeled graph as:

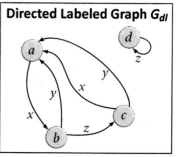

Directed Labeled Graph G_{dl}

$$G_{dl} = \{(a^b)^x, (b^a)^y, (b^c)^z, (c^a)^x, (c^a)^y, (d^d)^z\}.$$

We note that:

- G_{dl} is a relation, and each of its elements is a compound couplet.

- **yang** $(a^b)^x = x$ is the label and **yin** $(a^b)^x = a^b$ is the edge.

- By convention, the source (or origin) of the edge a^b is **yin** $a^b = a$, and the destination (or target) of the edge is **yang** $a^b = b$.

OK. We apologize in advance for the plethora of G's but "**G** stands for graph and "$\mathcal{G}$" stands for genesis set, so what can we do – invent a new word for "graph" or slip into a foreign language? Bearing the potential confusion in mind, we shall soldier on...

G_{dl} is a relation. If $\mathcal{G}$ is the genesis set for the elements a, b, c, x, y, etc., then we can also write:

$$G_{dl} \in \mathfrak{P}((\mathcal{G} \times \mathcal{G}) \times \mathcal{G}).$$

If this seems like a leap too far, remember back to when we wrote that Cartesian products create couplets. So the Cartesian product $\mathcal{G} \times \mathcal{G}$ can create couplets of the form a^b and thus the Cartesian product $((\mathcal{G} \times \mathcal{G}) \times \mathcal{G})$ can create couplets of the form $(a^b)^x$. The power set $\mathfrak{P}((\mathcal{G} \times \mathcal{G}) \times \mathcal{G})$ will therefore contain all possible relations (sets of couplets) of that form.

That in turn means that $\mathfrak{P}((\mathcal{G} \times \mathcal{G}) \times \mathcal{G})$ is a clan, because it is a set of relations. Indeed, it must be the set of all directed labeled graphs based on $\mathcal{G}$. So we can summarize by saying that:

$$\mathbb{G}_{dl} = \mathfrak{P}((\mathcal{G} \times \mathcal{G}) \times \mathcal{G}).$$

If we took our genesis set $\mathcal{G}$ to consist entirely of the Bush family tree and all possible familial relations, we would be able to create various representations of Bush Relations and various Bush clans. Although, given the terminology, it would probably be more fitting to involve the population of Scotland in that exercise.

Directed graphs

Now consider directed graphs with no labels. We can get to these by playing yin with G_{dl}:

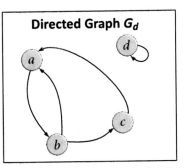

Directed Graph G_d

$$yin\ (G_{dl}) = \{a^b, b^a, b^c, c^a, d^d\}.$$

So the yin set of a directed labeled graph is, of course, the graph without the labels:

$$G_d = \{a^b, b^a, b^c, c^a, d^d\}.$$

This is a simpler situation than with directed labeled graphs: $G_d \in \mathfrak{P}(\mathcal{G} \times \mathcal{G})$ is just a common, garden variety relation.

An example of a graph (or network) of this kind is a map showing towns and the roads that connect them. A more interesting example, perhaps, is the map of the London Underground, which we will examine later.

Simple labeled graphs

We introduced the labeled graph with the Six Degrees of Kevin Bacon game. The movies were the labeled edges, the movie stars were the nodes and there was no direction (from star to star) involved.

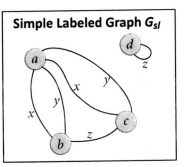

Simple Labeled Graph G_{sl}

We presented our solution to the Lee Marvin - Kevin Bacon example graphically, but we could also have represented it in this way:

{{*Lee Marvin, Donald Sutherland*}$^{Dirty\ Dozen}$, {*Donald Sutherland, Sir Ian McKellen*}$^{Six\ degrees\ of\ Separation}$, {*Sir Ian McKellen, Kevin Bacon*}$^{Xmen:\ First\ Class}$}.

Our genesis set $\mathcal{G}$ for this game is the set of all movies and movie actors. The cast of any movie is a subset of $\mathcal{G}$.

To play the Six Degrees of Kevin Bacon game, we need to pick pairs of actors, say:

$$\{Lee\ Marvin, Donald\ Sutherland\} \in \mathfrak{P}(\mathcal{G}),$$

to serve as undirected edges. However, if we want to go beyond this specific game, say to a game that included Tony Randall and the movie *7 Faces of Dr. Lao*, where Tony Randall plays seven different parts, we could encounter an undirected edge of the form:

$$\{\textit{Tony Randall, Tony Randall}\} = \{\textit{Tony Randall}\} \in \mathfrak{P}(\mathcal{G}),$$

indicating that in at least one movie, Tony Randall starred with Tony Randall.

Therefore, for the sake of generality, we need to identify that subset of $\mathfrak{P}(\mathcal{G})$ consisting of all subsets of $\mathcal{G}$ of cardinality one or two:

$$\mathfrak{p}(\mathcal{G}) := \{S \in \mathfrak{P}(\mathcal{G}) : 1 \leq |S| \leq 2\}.$$

It follows that the simple labeled graph of the Seven Degrees of Kevin Bacon game on page 104 and the simple labeled graph G_{sl}, indeed any simple labeled graph is an element of $\mathfrak{P}(\mathfrak{p}(\mathcal{G}) \times \mathcal{G})$.

Support: another function

We derived directed graphs from directed labeled graphs using the **yin** function, so you may be wondering whether we can also derive simple labeled graphs from directed labeled graphs. If not, then pause to wonder, then continue reading.

The answer is "yes," but the mild complication is that we need to invent another function for that specific purpose. So let's do that. Observe that the natural "flow" from G_{dl} to G_{sl} is characterized by the correspondence:

$$(a^b)^x \in \ \mathbf{G}_{dl} \longrightarrow \{a, b\}^x \in \ \mathbf{G}_{sl}.$$

Notice that, in this flow, the couplet a^b uniquely determines the set $\{a, b\}$ in the following way:

$$\{a, b\} = \{\textit{yin}(a^b), \textit{yang}(a^b)\}.$$

We can capture this mathematically by referring to $\{\textit{yin } a^b, \textit{yang } a^b\}$ as the **support** of a^b and writing:

$$\textit{spt}(a^b) := \{a, b\},$$

which provides us with a useful function. But sadly, it's only a halfway house. But let us not turn our nose up at it. It's useful, as you will realize later, hopefully.

It follows, by the way, that the support of a relation is the relation of supports (*the flarn of the clarp*... etc.). For example, notice that:

$$spt(G_d) = \{\{a, b\}, \{a, c\}, \{b, c\}, \{d, d\}\}$$

$$= \{\{a, b\}, \{a, c\}, \{b, c\}, \{d\}\}.$$

We will make use of this observation later. But let's get back to the natural flow from G_{dl} to G_{sl}: $(a^b)^x \longrightarrow \{a, b\}^x$. Clearly our **spt** function is involved here, but not "directly."

Notice in particular that:

$$spt((a^b)^x) = \{a^b, x\}.$$

So something a little more sophisticated is called for, which is this:

$$spt_{yin}((a^b)^x) := (spt(a^b))^x = \{a, b\}^x.$$

It follows that spt_{yin} is a function from $(\mathcal{G} \times \mathcal{G}) \times \mathcal{G}$ to $(\mathfrak{p}(\mathcal{G}) \times \mathcal{G})$. This function lifts naturally to $\mathfrak{P}(\mathcal{G} \times \mathcal{G}) \times \mathcal{G}$ and establishes that:

$$spt_{yin}(G_{dl}) = G_{sl}.$$

Simple graphs

We can think of simple graphs as "labeled graph without labels." Using our examples, that is:

$$yin\ (G_{sl}) = \{\{a, b\}, \{a, c\}, \{b, c\}, \{d\}\} = G_s$$

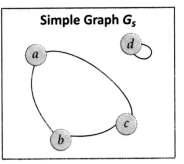

Simple Graph G_s

As we have just introduced the support function, it would be remiss not to draw attention to the fact that simple graphs can also be thought of as directed graphs without direction. So, we can write:

$$spt\ (G_l) = \{\{a, b\}, \{a, c\}, \{b, c\}, \{d\}\} = G_s.$$

This is fine as far as it goes, but it leaves us with a vaguely awkward situation: G_s is not a relation, and because all the other graphs are, we would much prefer that it was. How can we fix this and yet retain a mathematically faithful representation?

There is a way. Simply lift G_s to the diagonal of $G_s \times G_s$,

$$\{\{a, b\}^{\{a, b\}}, \{a, c\}^{\{a, c\}}, \{b, c\}^{\{b, c\}}, \{d\}^{\{d\}}\},$$

if the more cryptic G_s won't do.

Thus in general we can represent simple graphs as elements of:

$$\mathfrak{P}(\mathfrak{p}(\mathcal{G})) \text{ or } \mathfrak{P}(D_{\mathfrak{p}(\mathcal{G})})),$$

where $\mathfrak{p}(\mathcal{G}) := \{S \in \mathfrak{P}(\mathcal{G}) : 1 \leq |S| \leq 2 \}.$

Graphs and data

Graphs are important data structures because they are useful. A particularly good example of this is provided by maps. Maps were designed and used long before mathematicians dreamed up graph theory to systematize them. It turns out, as we hope you have gathered, that they are complex data structures.

For illustration, take a leisurely look at the example in Figure 10 on the next page – the map of The London Underground (or "the Tube," as Londoners prefer to call it). This map, of which we only show a portion, is generally regarded as a brilliant example of map design. The backstory is that Harry Beck, an engineering draftsman who worked for the London Underground, had a smouldering dislike for the geographically-correct map that London Underground displayed at all its stations.

He decided to create an alternative and in 1933 came up with a map of the form illustrated. A limited number of copies of his design were printed, and it became an instant success with the public and soon displaced the previous geographically-correct graph.

The London Underground was then, and still is, a complex interwoven system of different individual tube lines. Currently, there are 12 different Tube lines: The Bakerloo Line, The Central Line, The Circle Line, The District Line, The East London Line, The Hammersmith & City Line, The Jubilee Line, The Metropolitan Line, The Northern Line, The Piccadilly Line, The Victoria Line and The Waterloo & City Line. Each of these lines intersects with other lines at various stations, and they also intersect with the Docklands Light Railway and all the main railway stations that bring commuters into London.

Beck realized that in order to navigate the Tube, the traveler's most important priority was what lines stations were on and how these lines

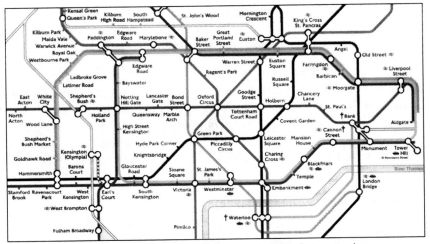

Figure 10: The map of the London Underground

were connected. He reasoned that actual geography didn't matter much and rather than be a slave to it, he placed the stations roughly evenly along lines which he ran either horizontally, vertically or at 45 degrees. He probably never thought about it in those terms, but it was an exercise in topology.

The outcome was that he shrank the map to fit it into a much smaller space with a design that was much easier to read. By squeezing and stretching the tube lines and even the River Thames itself, Beck produced a design that was eventually copied by many transport systems around the world.

Not only did this design stand the test of time, but as it evolved, more and more detail was added. Using various symbols and graphical nuances, tube stations on the map now show interchange points, disabled access, connection to river transport, connection to rail transport, airport connections and those stations closed on Sundays. The different tube lines (the edges of this graph) are color coded, although as we are limited to a black and white illustration, this is only apparent by the different shades of grey used for the tube lines.

As I'm sure you have worked out, it qualifies as a simple labeled graph. If we take just a small central portion of the graph that includes Bond Street, Oxford Circus, Green Park and Piccadilly Circus, we can represent it with the following relation:

$\{\{$*Bond Street, Oxford Circus*$\}^{Central}$, $\{$*Bond Street, Green Park*$\}^{Jubilee}$, $\{$*Green Park, Oxford Circus*$\}^{Victoria}$, $\{$*Green Park, Piccadilly Circus*$\}^{Piccadilly}$, $\{$*Oxford Circus, Piccadilly Circus*$\}^{Bakerloo}\}$

In this relation, the yang of each couplet (i.e., the edge) is one of the tube lines: Central, Jubilee, Victoria or Bakerloo. The yin of the couplet is the set of cardinality two containing the two tube station names.

Imagine that you are planning to go from Bond Street to Piccadilly Circus. To do so you will need to make a mental map of the route which will involve you changing from one tube line to another. There are a number of ways that this can be achieved. One route would be to go from Bond Street to Oxford Circus on the Central Line, then change to the Victoria Line to go from Oxford Circus to Green Park, and then change onto the Piccadilly Line to go to Piccadilly Circus.

This route is a directed labeled graph:

$$\{(Bond\ Street^{Oxford\ Circus})^{Central}, (Oxford\ Circus^{Green\ Park})^{Victoria},$$
$$(Green\ Park^{Piccadilly\ Circus})^{Piccadilly}\},$$

as indeed are all other possible routes.

The yin of this route:

$$\{Bond\ Street^{Oxford\ Circus}, Oxford\ Circus^{Green\ Park}, Green\ Park^{Piccadilly\ Circus}\},$$

marks out a "path."

Graphs and semantics

Graphs can also be used in a very different context, to store the meaning of language as a collection of nodes and edges. This is becoming their most important area of application. All textual data in any language you please can be represented using directed graphs. Consider Figure 11, on the next page, which shows most of the lyrics to *Hey Jude* represented as a directed graph.

If, for example, you just wish to follow the final repeating refrain, "Na, Na, Na, Na, Na, Na, Na, Na, Na, Na, Na, Hey Jude," you can do so by following the graph, starting at the node Na. The set of couplets for this refrain is:

$$\{Na^{Na}, Na^{Hey\ Jude}\}.$$

The problem with this is that it does not tell you how many times to sing "Na." So, what you might prefer is an ordered 12-tuple, like this:

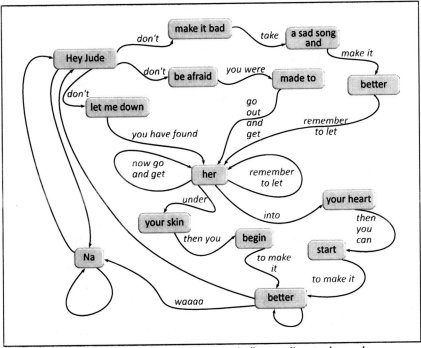

Figure 11: The lyrics to "Hey Jude," as a directed graph.

$$\{(Na^{Na})^1, (Na^{Na})^2, (Na^{Na})^3, (Na^{Na})^4, (Na^{Na})^5, (Na^{Na})^6, (Na^{Na})^7, (Na^{Na})^8,$$
$$(Na^{Na})^9, (Na^{Na})^{10}, (Na^{Na})^{11}, (Na^{Hey\ Jude})^{12}\}$$

to indicate the order and repetitions of "Na."

Incidentally, if the goal is to reflect the semantics of language, Figure 11 is not a good application of a directed graph. If you think in simple grammatical terms, words target other words, and words are collected together into phrases. So adjectives, prepositions and articles target nouns and adverbs target verbs. Most sentences in a language break down into triples of the form subject-predicate-object, for example:

"Data Algebra (subject) haunts (predicate) your dreams (object)."

We could write this as a triple in the following way:

$$\{Data\ Algebra^{subject}, haunts^{predicate}, your\ dreams^{object}\}.$$

If we wanted to get a little more refined and break it down into individual words, we could also represent "Data Algebra" as the couplet: $Algebra^{Data}$

and "your dreams" as the couplet $dreams^{your}$. We would then have the following triple:

$$\{(Algebra^{Data})^{subject}, haunts^{predicate}, (dreams^{your})^{object}\}.$$

The subject of a sentence doesn't have to have an object. The object can be implied as in "Data Algebra (subject) rules (predicate)." We might be tempted to write that as:

$$\{(Algebra^{Data})^{subject}, rules^{predicate}\}.$$

Having demonstrated, if only in a simple way, that we can represent textual data with directed graphs, it's probably a good time for us to introduce the topic of ontologies. If you don't know what an ontology is, shame on you! Look it up in a dictionary, and you'll be even more mystified. The dictionary will probably tell you something along the lines of:

"Ontology is the branch of metaphysics that deals with the nature of being."

This should have you wondering why the word would ever need to be plural. That definition used to be fine until Artificial Intelligence (AI) aficionados started to try to make sense of human language and needed to build semantic structures for the purpose of studying and analyzing specific "domains."

In that area of activity, an ontology is a formal, standardized explicit description of concepts (classes) in a specific domain, with properties of each concept describing associated features, attributes and constraints. Combine an ontology with a set of individual instances of classes, and you have what could be called a knowledge base. If you are not confused by that, either you're very smart or you've dozed off. Instead, think of it like this: in an area such as, say, pharmaceuticals, an ontology defines a formal semantic framework that eliminates ambiguity and enables analysis.

As you probably suspected all along, ontologies can be defined as triples.

From couplets to relations to clans (diamonds are forever)

It's time now to return to talking about graphs in a formal mathematical manner. Before we took a sight-seeing tour round the world of semantics, by way of the London Underground, we had described each of the four different kinds of graph. In each case we came to the conclusion that the graph could be represented precisely by a relation. So our different examples of graphs, G_{dl}, G_d, G_{sl} and G_s are relations.

It naturally follows that the set of all graphs of each particular kind (directed labeled, directed, simple labeled and simple) drawn from a particular genesis set $\mathcal{G}$ are clans. We can write these clans as follows:

The clan of directed labeled graphs:

$$G_{dl} = \mathfrak{P}((\mathcal{G} \times \mathcal{G}) \times \mathcal{G}).$$

The clan of directed graphs:

$$G_d = \mathfrak{P}(\mathcal{G} \times \mathcal{G}) = \text{Relations}(\mathcal{G}).$$

The clan of simple labeled graphs:

$$G_{sl} = \mathfrak{P}(\mathfrak{p}(\mathcal{G}) \times \mathcal{G}),$$

$$\mathfrak{p}(\mathcal{G}) := \{S \in \mathfrak{P}(\mathcal{G}) : 1 \leq |S| \leq 2 \}.$$

The clan of simple graphs,

$$G_s = \mathfrak{P}(D_{\mathfrak{p}(\mathcal{G})}),$$

$$\mathfrak{p}(\mathcal{G}) := \{S \in \mathfrak{P}(\mathcal{G}) : 1 \leq |S| \leq 2 \}.$$

You will have noticed, as we moved through the description of each of the four kinds of graph, that it was possible to derive the form of the elements of the less complex graphs from the elements of the more complex ones. We illustrate this in Figure 12, below.

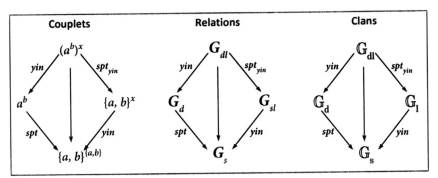

Figure 12: A display of diamonds

You can also think of this as mathematical calisthenics, if you like. You can manipulate these relations in the same dextrous way that you manipulated other algebraic symbols and operators in your youth.

An alternative representation

One useful point to note about graphs is that there is an alternative set-theoretical representation of them. We shall conclude this chapter by explaining it.

The relation, $G_{dl} = \{(a^b)^x, (b^a)^y, (b^c)^z, (c^a)^x, (c^a)^y, (d^d)^z\}$ can also be represented as a clan. Here is how:

$$G_{dl} = \{\{\, a^1, b^2, x^3\},$$
$$\{\, b^1, a^2, y^3\},$$
$$\{\, b^1, c^2, z^3\},$$
$$\{\, c^1, a^2, x^3\},$$
$$\{\, c^1, a^2, y^3\},$$
$$\{\, d^1, d^2, z^3\}\}.$$

We have taken the couplets $(a^b)^x$, etc., and represented them as ordered 3-tuples. There is a one-to-one correspondence between the couplets in the relation G_{dl} and the relations in the clan G_{dl}.

This means that we can go from one algebraic structure to the other and back, if need be. We can take some G_{dl} and transform it into a clan, G_{dl}, or take a clan G_{dl} and transform into a relation G_{dl}. This is more mathematical calisthenics.

If you are wondering whether we can play the same trick with all the other kinds of graphs, then cease to wonder. We can. The clans that emerge are slightly different in each case, as you might imagine. And if you want to find out what they are, you'll have to work it out.

In the time-honored tradition of authors of mathematics books, we have decided to leave it as an exercise for the reader.

Taking Stock

This chapter climbs the algebraic tower to consider the algebra Clans($\mathcal{G}$) and the lift of operators from Relations($\mathcal{G}$) to Clans($\mathcal{G}$). The following bullet points summarize its content:

- A clan is a set of relations.

- The operations of transposition and composition lift from Relations($\mathcal{G}$) to Clans($\mathcal{G}$) in a natural way.

- The yin-set of a clan $\mathbb{C}$, $yin(\mathbb{C})$, is the union of the yin sets of the relations in $\mathbb{C}$. We can write this:

$$yin(\mathbb{C}) = \cup_{R \in \mathbb{C}} \; yin(R).$$

  Similarly:

$$yang(\mathbb{C}) = \cup_{R \in \mathbb{C}} \; yang(R).$$

- In Clans($\mathcal{G}$), the union $\cup$ and intersection $\cap$ operators both apply, because it is a set algebra. However, we defined two further operators: the **cross-union** and the **cross-intersection** to offer alternative forms of union and intersection. They are defined as follows:

  The cross-union of two sets of sets is the set of all possible unions of sets from the original sets and is denoted by ▼.

  The cross-intersection of two sets of sets is the set of all possible intersections of sets from the original sets and is denoted by ▲.

- We identified four distinct types of graph: the directed labeled graph, the directed graph, the simple labeled graph and the simple graph. All of these representations of data can be defined as relations. Examples are as follows:

  Directed labeled graph: $G_{dl} = \{(a^b)^x, (b^a)^y, (b^c)^z, (c^a)^x, (c^a)^y, (d^d)^z\}$,

  Directed graph: $G_d = \{a^b, b^a, b^c, c^a, d^d\}$,

  Simple labeled graph: $G_{sl} = \{\{a, b\}^x, \{a, c\}^x, \{a, b\}^y, \{a, c\}^y, \dots \{d\}^z\}$,

  Simple graph: $G_s = \{\{a, b\}^{\{a, b\}}, \{a, c\}^{\{a, c\}}, \{b, c\}^{\{b, c\}}, \dots \{d\}^{\{d\}}\}$.

- Such graphs can also be represented as clans. An example is:

$$\mathbf{G_{dl}} = \{\{\,a^1, b^2, x^3\},$$
$$\{\,b^1, a^2, y^3\},$$
$$\{\,b^1, c^2, z^3\},$$
$$\{\,c^1, a^2, x^3\},$$
$$\{\,c^1, a^2, y^3\},$$
$$\{\,d^1, d^2, z^3\}\}$$

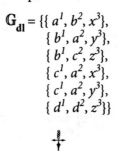

Chapter 8: Joinery

It is easier to square the circle than to get round a mathematician.

~ De Morgan

———⚌———

SQL and the axiom of specification

You may have thought that you left all those set theory axioms behind in Chapter 3, but you didn't. One of them has returned to sit on your knee; its the axiom of specification. This statement of the crashingly obvious proclaims that: *Every set* **A** *and every statement* **S(x)** *determines a unique subset* **S** *of* **A** *whose elements are exactly those elements* **x** *of* **A** *for which* **S(x)** *holds.*

OK, we realize the axiom's formal statement may not be aesthetically pleasing, but all it really says is that sets can have subsets, and you can get a subset by making some statement or other about elements of that set to create the subset. When it comes to real data stored in a computer, one of the ways of making such a statement is using a SQL **SELECT**.

If you are reading this book and you don't know what SQL is, we have to confess that we're surprised. But, as we are about to explore SQL, we shall allow for that possibility and explain what it is and we shall also explain every SQL statement[6] we discuss.

SQL (often pronounced "sequel") is a special-purpose language designed for accessing and managing data held in relational databases. You could call it a programming language, but it is not a general purpose programming language like C++, Java or Python. It's capabilities are divided between data definition (defining data structures) and data manipulation (accessing and manipulating the data).

You also need to know that there is a SQL standard, the first version of which was fathered by the American National Standards Institute (ANSI) in 1986 and adopted by the International Organization for Standardization

---

6   In practice, this means that the data, no matter how it is actually stored, is logically defined by a schema that represents the data as if it were stored in regular well-described tables. For convenience, in this chapter we will visually represent data in tables.

(ISO) in 1987. Since then it has regularly been updated, and vendors of relational database products have tended to abide by it with minor variances here and there. In the SQL examples that follow, we abide by the version that Microsoft implements.[7]

The most commonly used statement in SQL is the **SELECT** statement. It enables the user or programmer to **SELECT** data from tables – in effect, taking a subset of a table of data held in the database. However, before we examine that, we'll use SQL to create a table.

The **CREATE TABLE** SQL syntax is as follows:

CREATE TABLE *table_name*
   (
   column_name1 data_type(size),
   column_name2 data_type(size),
   column_name3 data_type(size),
   ....
   );

The *column_name* parameters name the table columns. The *data_type* parameter specifies the column's data type (e.g., varchar, integer, date, etc.). The size parameter specifies the maximum length of the table column.

For example:

CREATE TABLE *Outlaws*
   (
   ID int,
   Name varchar(255),
   Crime varchar(255),
   Reward varchar(255),
   BornIn varchar(255),
   LastSeen varchar(255)
   );

If we execute this SQL statement, the outcome will be to create an empty table in a database and to update the database schema (the database's formal description of the data structures it holds). The equivalent algebraic form, if we were going to read all or part of that table, would be to record the table's data structure as a relation, which we can name *OL_head*:

---

7   *This does not indicate a particular preference; it was simply more convenient as we used a Microsoft SQL reference as a source.*

$OL\_head = \{ID^{int}, Name^{v1}, Crime^{v1}, Reward^{v1}, BornIn^{v1}, LastSeen^{v1}\}$.

Note that above we have used the value v1 as a shorthand for the couplet varchar^{255}, where $yin(v1)$ = varchar and $yang(v1)$ = 255. Most databases support a variety of data types, some of which are couplets. So in general, if we are dealing with a relation R that defines a table row, $yang$ (R) will be a collection of data types which may or may not itself be a relation. In this instance:

$$yang\,(OL\_header) = \{int, varchar^{255}\},$$

which is not a relation – it is merely a set with one element that is a couplet. If you'd like a visual representation of this table header, then maybe this will satisfy you:

Table 3: Outlaws table structure					
ID	Name	Crime	Reward	BornIn	LastSeen
Int	varchar255	varchar255	varchar255	varchar255	varchar255

SQL users who are requesting data directly have little interest in the data types we show here, because their software manages all of that physical detail. They are interested in the data values that make up the rows of the table and the header elements which they understand as data descriptions.

Simple SQL select statements

If we assume that the table we created with our CREATE TABLE statement has just had some data poured into it, we might use a simple SELECT statement to retrieve some of that data. The SELECT statement has a variety of forms, the simplest of which only retrieve data from a single table. We will examine these first.

The following request asks for a number of specific columns of data to be retrieved from a table:

SELECT *column_name1, column_name2,...* FROM *table_name*;

Here's an example that queries the **Outlaws** table (see next page):

SELECT *Name, Crime, LastSeen*
FROM *Outlaws;*

The values in the table are as shown:

121

Table 4: Outlaws table					
ID	Name	Crime	Reward	BornIn	LastSeen
1	Robin Hood	Larceny	100 guineas	Locksley	Sherwood Forest
2	Billy the Kid	Murder	$500	New York	On the lam
3	Al Capone	Tax Evasion	None	New York	Frank Nitti's speakeasy
4	Loki	Deception	Thor's gratitude	Asgard	On Earth goddamit!
5	Chewbacca	Treason	Bounty	Kashyyyk	Tatouine

Take care to note that the table illustrated above and other tables shown in the rest of this chapter are merely convenient artifices to represent clans in a similar form to the form that relational databases represent the same data. So, from that perspective, the table above is a convenient visual artifice for the clan:

Outlaws =

$\{\{1^{ID},$ *Robin Hood*$^{Name},$ *Larceny*$^{Crime},$ *100 guineas*$^{Reward},$ *Locksley*$^{BornIn},$ *Sherwood Forest*$^{LastSeen}\},$
$\{2^{ID},$ *Billy The Kid*$^{Name},$ *Murder*$^{Crime},$ *$500*$^{Reward},$ *New York*$^{BornIn},$ *On the lam*$^{LastSeen}\},$
$\{3^{ID},$ *Al Capone*$^{Name},$ *Tax Evasion*$^{Crime},$ *None*$^{Reward},$ *New York*$^{BornIn},$ *Frank Nitti's speakeasy*$^{LastSeen}\},$
$\{4^{ID},$ *Loki*$^{Name},$ *Deception*$^{Crime},$ *Thor's Gratitude*$^{Reward},$ *Asgrad*$^{BornIn},$ *On Earth goddamit!*$^{LastSeen}\},$
$\{5^{ID},$ *Chewbacca*$^{Name},$ *Treason*$^{Crime},$ *Bounty*$^{Reward},$ *Kashyyyk*$^{BornIn},$ *Tatouine*$^{LastSeen}\}\}.$

We perform the algebraic equivalent of the SQL query: **SELECT** *Name, Crime, LastSeen* **FROM** *Outlaws;* with a composition in the following way:

$\mathbb{C}$ = Outlaws ∘ $\{\{Name^{Name}, Crime^{Crime}, LastSeen^{LastSeen}\}\}$

$= \{\{$*Robin Hood*$^{Name},$ *Larceny*$^{Crime},$ *Sherwood Forest*$^{LastSeen}\},$
$\{$*Billy The Kid*$^{Name},$ *Murder*$^{Crime},$ *On the lam*$^{LastSeen}\},$
$\{$*Al Capone*$^{Name},$ *Tax Evasion*$^{Crime},$ *Frank Nitti's speakeasy*$^{LastSeen}\},$
$\{$*Loki*$^{Name},$ *Deception*$^{Crime},$ *On Earth goddamit!*$^{LastSeen}\},$
$\{$*Chewbacca*$^{Name},$ *Treason*$^{Crime},$ *Tatouine*$^{LastSeen}\}\}.$

Consider just the composition of the first relation in the first clan with the only relation in the second clan:

$\{1^{ID}$, *Robin Hood*^{Name}, *Larceny*^{Crime}, *100 guineas*^{Reward}, *Locksley*^{BornIn}, *Sherwood Forest*$^{LastSeen}\}$ $\circ\{Name^{Name}$, *Crime*^{Crime}, *LastSeen*$^{LastSeen}\}$
$= \{Robin Hood^{Name}$, *Larceny*^{Crime}, *Sherwood Forest*$^{LastSeen}\}$.

So the result shown emerges when we apply that composition to every relation of **Outlaws**.

Incidentally, it is worth noting that the SQL **SELECT** would normally be used to retrieve data from a database, but the examples we are giving here are of querying tables held in memory as clans. The result that the SQL statement would usually furnish would just provide the values in the rows, with the metadata (the table heading) being implicit.

A simple variant of the SQL **SELECT** is **SELECT DISTINCT**. It has the following form:

SELECT DISTINCT *column_name1, column_name2,...* **FROM** *table_name;*

Consider the following example.

    SELECT DISTINCT *BornIn*
    FROM *Outlaws;*

Again, we perform the algebraic equivalent of the SQL query with a composition in the following way:

Outlaws $\circ$ $\{\{BornIn^{BornIn}\}\} = \{\{Locksley^{BornIn}\}$,
                                  $\{New York^{BornIn}\}$,
                                    $\{Asgard^{BornIn}\}$,
                                    $\{Kashyyyk^{BornIn}\}\}$.

Two of the rows in the **Outlaws** table have the value "New York" under the BornIn heading. These are reduced to a single value by the composition.

Introducing Some Completely New Operators!

SQL provides a **WHERE** clause to impose conditions on the query. This is one example of this form of **SELECT**:

SELECT * **FROM** *table_name* **WHERE** *column_name1 operator value* **OR** *column_name2 operator value;*

The **SELECT** * means "select all that qualify." The logical operators **AND** and **OR** can be used and any number of different columns can be used to qualify the query.

Here is an example of this form:

SELECT * **FROM** *Outlaws* **WHERE** *Crime = 'Murder'* **OR** *Crime = 'Treason';*

Believe it or not, to get to these answers we need to introduce a wholly new operator that looks like this: $\rhd$, which goes by the name of ***superstriction***. As you might have suspected, it has a partner operator denoted by $\lhd$ and calls itself ***substriction***. That isn't all we need to do. But let's begin there.

For Relations($\mathcal{G}$) we define ***superstriction*** as follows:

$$A \rhd B := A \text{ if } A \supset B,$$

and we define ***substriction*** as follows:

$$A \lhd B := A \text{ if } A \subset B.$$

Take note: these are partial operators. They give a result only if one set is a subset of the other. To be specific, superstriction produces a result only if the relation to the right of the operator is a subset of the one to the left, otherwise the outcome is undefined. And substriction produces a result only if the relation to the left of the operator is a subset of the one to the right, otherwise the outcome is undefined. Consequently, both of these operators are partial operators.

It will probably be easier to see what this means using an example. So consider the relations below:

$A_1 = \{yes^{yes}, no^{no}, maybe^{yes}, perhaps^{yes}, perchance^{yes}\}.$
$B_1 = \{yes^{yes}, maybe^{yes}, perhaps^{yes}, perchance^{yes}\}.$
$B_2 = \{no^{no}, perchance^{no}, I \text{ seriously doubt it}^{maybe}\}.$
$B_3 = \{yes^{yes}, no^{no}, maybe^{yes}, perhaps^{yes}, perchance^{yes}, who \text{ knows}^{not\ I}\}.$

Applying the superstriction:

$$A_1 \rhd B_1 = A_1,$$

which is the case because B_1 is clearly a subset of A_1. Also:

$$A_1 \rhd B_2$$

is undefined, because, although B_2 contains a couplet that is also found in A_1, it has couplets that are not.

$$A_1 \rhd B_3$$

is also undefined because B_3 contains all the couplets of A_1 but also an extra couplet that disqualifies it from providing a result.

Now peruse our next set of example relations:

$A_2 = \{kernel^{yes}, seed^{yes}, root^{yes}, germ^{yes}\}.$
$B_4 = \{kernel^{yes}, seed^{yes}, root^{yes}, germ^{yes}, who\ cares^{yes}\}.$
$B_5 = \{seed^{yes}, root^{yes}, germ^{yes}\}.$
$B_6 = \{kernel^{yes}, seed^{yes}, root^{yes}, it\ don't^{fit}\}.$

The following are examples of substrictions:

$$A_2 \lhd B_4 = A_2,$$

because A_2 is clearly a subset of B_4.

$$A_2 \lhd B_5$$

is undefined, because B_5 is a subset rather than a superset of A_2.

$$A_2 \lhd B_6$$

is undefined, because B_6 has an element that is not found A_2.

Now that we have defined our brand new substriction and superstriction operators, we need to lift them to Clans($\mathcal{G}$). Yes, indeed. We created a new pair of operators lower down in the tower of data algebra just so we could lift them. And the really cool thing about lifting them is that we don't get all those troublesome undefined results. But it isn't plain sailing.

Why? Well, as you should know by now: *the-flarn-of-the-clarp-is-very-often-the-clarp-of-the-flarns*. And that's how it works out in this case. The superstriction of Clans is the Clan of superstrictions. Similarly, the substriction of Clans is the Clan of substrictions.

And, while we're at it, we will need a couple of precise definitions to hang on these operators. So let's define the cross-superstriction:

> *The cross-superstriction of two clans is the clan of all the superstrictions between the relations in the original clans, and is denoted by* ▶.

Similarly:

> *The cross-substriction of two clans is the clan of all the substrictions between the relations in the original clans, and is denoted by* ◀.

Let us provide examples so you can see what's going on. Consider clans $\mathbb{A}$ and $\mathbb{B}$, where:

$$\mathbb{A} = \{\{Billy\ The\ Kid^{Name},\ Murder^{Crime},\ \$500^{Reward}\},$$
$$\{Chewbacca^{Name},\ Treason^{Crime},\ Bounty^{Reward}\},$$
$$\{Jack\ Sparrow^{Name},\ Piracy^{Crime},\ \$1000\ guineas^{Reward}\},$$
$$\{Mata\ Hari^{Name},\ Espionage^{Crime},\ untold\ secrets^{Reward}\}\}$$

and:

$$\mathbb{B} = \{\{Jack\ Sparrow^{Name},\ Piracy^{Crime}\},$$
$$\{Mata\ Hari^{Name},\ Espionage^{Crime}\},$$
$$\{Bernie\ Madoff^{Name},\ Embezzlement^{Crime}\}\}.$$

The cross-superstriction of the two clans, $\mathbb{A}$ and $\mathbb{B}$ is:

$$\mathbb{A} ▶ \mathbb{B} = \{\{Jack\ Sparrow^{Name},\ Piracy^{Crime},\ \$1000\ guineas^{Reward}\},$$
$$\{Mata\ Hari^{Name},\ Espionage^{Crime},\ untold\ secrets^{Reward}\}\}.$$

And, the cross-substriction of the two clans $\mathbb{B}$ and $\mathbb{A}$ is:

$$\mathbb{B} ◀ \mathbb{A} = \{\{Jack\ Sparrow^{Name},\ Piracy^{Crime}\},$$
$$\{Mata\ Hari^{Name},\ Espionage^{Crime}\}\}.$$

If you remember rightly, a few pages back we set out trying to answer a SQL query and discovered that our beloved data algebra was short of an operator or two or three or four. Trying desperately not to admit that to you, the reader, we hastily descended the tower created some convenient operators to help us out, and then we came back.

This is the tried and true mathematical tactic of: "take your problem some place else, do something there and then come back." We shimmied down the tower to Relations($\mathcal{G}$), created some new operators, put them in a bag, then clawed our way back up into Clans($\mathcal{G}$) where, after a little cross-pollination, we had what we wanted. It only remains for us to demonstrate that these operators are indeed what we need. So let's now take a swing at that SQL **SELECT** statement that caused all the trouble:

SELECT * **FROM** *Outlaws* **WHERE** *Crime* = 'Murder' **OR** *Crime* = 'Treason';

We calculate the required clan $\mathbb{C}$ in the following way:

$$\mathbb{C} = \text{Outlaws} \blacktriangleright \{\{\text{Murder}^{Crime}\}, \{\text{Treason}^{Crime}\}\}$$
$$= \{\{2^{ID}, \textit{Billy The Kid}^{Name}, \textit{Murder}^{Crime}, \$500^{Reward}, \textit{New York}^{BornIn},$$
$$\textit{On the lam}^{LastSeen}\},$$
$$\{5^{ID}, \textit{Chewbacca}^{Name}, \textit{Treason}^{Crime}, \textit{Bounty}^{Reward}, \textit{Kashyyyk}^{BornIn},$$
$$\textit{Tatouine}^{\ LastSeen}\}\}.$$

If you take another look at the **Outlaws** table (page 122), you can see that this clan provides the equivalent result to performing the above SQL **SELECT** on that table, furnishing the table rows for that heinous murderer Billy The Kid and the ill-tempered treasonous Chewbacca.

We can now change the subject and write a few words about SQL operators.

SQL operators

The SQL *operator* in the SQL statement:

SELECT * **FROM** *table_name* **WHERE** *column_name1 operator value* **OR** *column_name2 operator value;*

can be any one of those shown in the adjacent table.

Note that some of the operators shown (<, >, etc.) presume that the values in a column have a natural order, which may not be the case. In practice, specific database products have specific rules for sorting data (collation rules), which they automatically apply to all SQL requests.

Table 5: SQL Operators	
Operator	**Description**
=	Equal
<>	Not equal
>	Greater than
<	Less than
>=	Greater than or equal
<=	Less than or equal
BETWEEN	Between an inclusive range
LIKE	Search for a pattern
IN	To specify a set of multiple possible values

If we know that there is a valid ordering for a given table column, we can easily construct a corresponding algebraic equation that parallels SQL statements of this kind. An example might be as follows:

SELECT * FROM *Outlaws* **WHERE** *BornIn* **BETWEEN** *"Asgard"* **AND** *"Kashyyyk"*;

We can evaluate this with the following steps. First we evaluate:

$$\mathbb{C}_1 = \textbf{Outlaws} \circ \{\{BornIn^{BornIn}\}\}$$
$$= \{\{Locksley^{BornIn}\},$$
$$\{New\ York^{BornIn}\},$$
$$\{Asgard^{BornIn}\},$$
$$\{Kashyyyk^{BornIn}\}\}$$

Then we create:

$$\mathbb{C}_2 := \{c^{BornIn} \in \mathbb{C}_2 : c \geq \text{"Asgard" and } c \leq \text{"Kashyyyk"}\}$$
$$= \{\{Asgard^{BornIn}\}, \{Kashyyyk^{BornIn}\}\}$$

We then evaluate:

$$\mathbb{C}_3 = \textbf{Outlaws} \blacktriangleright \mathbb{C}_2$$

$$= \textbf{Outlaws} \blacktriangleright \{\{Asgard^{BornIn}\}, \{Kashyyyk^{BornIn}\}\}$$

$$= \{\{4^{ID}, Loki^{Name}, Deception^{Crime}, Thor's\ Gratitude^{Reward}, Asgrad^{BornIn},$$
$$On\ Earth\ goddamit!^{LastSeen}\},$$
$$\{5^{ID}, Chewbacca^{Name}, Treason^{Crime}, Bounty^{Reward}, Kashyyyk^{BornIn},$$
$$Tatouine^{LastSeen}\}\}.$$

Keys

Before we move on to discussing **JOINs**, we'd like to say a few words about keys, since in some circumstances (e.g., for a primary key) it is useful to designate table columns or combinations of columns as keys. So, to illustrate this, let's examine the Hunter table shown on the next page, comprising a set of individuals driven by personal motives to pursue particular targets.

The situation may arise where we want to create a unique key for this table. Given how few elements this table has, we can work out some possible solutions just by looking. With much larger clans, including clans that do not fit conveniently into tables, we require an algebraic approach that covers all possibilities.

ID	Name	Title	Target	Motive
	Table 6: Hunter table			
a	Pat Garrett	Sheriff	Billy the Kid	$500 reward
b	Bellerophon	Slayer	Chimera	Stheneboea's hand in marriage
c	Boba Fett	Bounty Hunter	Chewbacca	Bounty (not known)
c	Boba Fett	Bounty Hunter	Han Solo	Dislike
d	Elliot Ness	Chief Investigator	Al Capone	Dislike

So consider the diagonal $D_{yang(\mathbb{H})}$ of the Cartesian product of $yang(\mathbb{H})$ with itself:

$$D_{yang(\mathbb{H})} = \{ID^{ID}, Name^{Name}, Title^{Title}, Target^{Target}, Motive^{Motive}\}.$$

Any subset K of $D_{yang(\mathbb{H})}$ where $|(\mathbb{H} \circ \{K\})| = |\mathbb{H}|$ is referred to as a **key** for $\mathbb{H}$.

In simple words, "a key is a subset of the diagonal where the result of its composition with $\mathbb{H}$ has the same cardinality as $\mathbb{H}$."

$D_{yang(\mathbb{H})}$ in this example has lots of subsets – 30, if you exclude the relation itself and the empty set Ø. Some of them, $\{Name^{Name}\}$ or $\{Motive^{Motive}\}$ for example, cannot be keys:

$$\mathbb{H} \circ \{Name^{Name}\} = \{\{Pat\ Garrett^{Name}\}, \{Bellerophon^{Name}\}, \{Boba\ Fett^{Name}\}, \{Elliot\ Ness^{Name}\}\}.$$

So:

$$|\mathbb{H} \circ \{Name^{Name}\}| = 4, \text{ while } |\mathbb{H}| = 5.$$

Similarly,:

$$|\mathbb{H} \circ \{Motive^{Motive}\}| = 4, \text{ while } |\mathbb{H}| = 5.$$

Moreover the relation $\{Name^{Name}, Motive^{Motive}\}$ is also a key, because:

$$|\mathbb{H} \circ \{Name^{Name}, Motive^{Motive}\}| = 5 = |\mathbb{H}|.$$

However, it may not be the most efficient key. That might be the relation $\{Target^{Target}\}$. If you check you'll see that:

$$|\mathbb{H} \circ \{Target^{Target}\}| = 5 = |\mathbb{H}|.$$

And this, as a relation, has the least possible cardinality for a key: one.

A *minimal key* for a clan is defined to be a key of least cardinality, and the minimal key for $\mathbb{H}$ is clearly $\{Target^{Target}\}$. A minimal key is likely to be the preferred choice for a key.

The SQL JOIN Family

The SQL **SELECT** statements that we have discussed so far were all directed against a single table. There are other forms of the **SELECT** statement that **JOIN** the data from the two tables together using the values from a column in each table. Tables *X* and *Y* below provide a simple example, showing the result:

Table 7: Table X		
noun	**adjective**	**enjoin**
good	kosher	join me
mismatch	unsuitable	ignore me

Table 8: Table Y		
verb	**adverb**	**enjoin**
fit	nicely	join me
reject	quickly	shun me

Table 9: JOIN of X with Y using the "enjoin" column				
enjoin	**noun**	**adjective**	**verb**	**adverb**
join me	good	kosher	fit	nicely

The way the **JOIN** illustrated above works is that the row from table *X* that has the same value in the *enjoin* column as the row from table *Y*, "*join me*," is united with that row to create a new table that has just a single row. In the language of data algebra, we could represent this result as:

$$\{good^{noun}, kosher^{adjective}, join\ me^{enjoin}\} \cup \{fit^{verb}, nicely^{adverb}, join\ me^{enjoin}\}$$
$$= \{good^{noun}, kosher^{adjective}, fit^{verb}, nicely^{adverb}, join\ me^{enjoin}\}$$

Clearly **JOIN**s are selective unions of a kind. There are many different varieties of **JOIN**, many of which, such as theta joins, are elaborations of simpler **JOIN**s. We will not discuss those here. Our immediate goal is merely to demonstrate that we can do something equivalent to a SQL **JOIN** using data algebra.

In standard SQL there are seven types of **JOIN**:

- The **INNER JOIN**
- The **LEFT JOIN**
- A version of the **LEFT JOIN** that is exclusive
- The **RIGHT JOIN**
- A version of the **RIGHT JOIN** that is exclusive
- The **OUTER JOIN**
- A version of the **OUTER JOIN** that is exclusive.

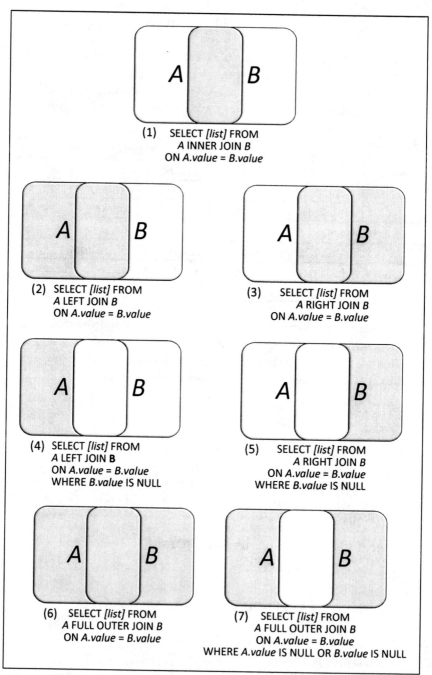

Figure 13: All JOINs

We illustrate all of these in Figure 13, along with the corresponding SQL **SELECT** statements. It has become common in SQL texts to use Venn diagrams to represent the different JOINs. As this is misleading (a **JOIN** is not a corresponding set operation to any of the set theory operators that Venn diagrams represent), we've chosen to use a slightly different visual artifice, preferring rounded squares to circles.

We will use the following two small tables to illustrate the seven different JOINs.

Table 10: Example table A	
Num	A_Words
1	dead
6	unexploded
3	joke
5	spam
10	Spanish Inquisition

Table 11: Example table B	
Num	B_Words
9	Ron Obvious
8	silly walks
1	parrot
3	warfare
6	Scotsman

A SQL **INNER JOIN** statement takes the following general form:

SELECT *[list]* **FROM** *table_name1* **INNER JOIN** *table_name2* **ON** *table_name1.value = table_name2.value;*

You can also add a **WHERE** clause to the **SELECT**, but for the sake of simplicity we will not do that for the **INNER JOIN**. As can be seen from Figure 13, other types of SQL **JOIN** involve a **WHERE** statement.

An **INNER JOIN** on the tables *A* and *B* could be written as follows:

SELECT * **FROM** *A* **INNER JOIN** *B* **ON** *A.Num = B.Num;*

Table 12: INNER JOIN of table A and B		
Num	A_Words	B_Words
1	dead	parrot
6	unexploded	Scotsman
3	joke	warfare

The result, shown in the table above is as you might expect. It joins data taken from rows of table *A* and *B* that have the same value for *Num*.

If we now consider this algebraically, we have two clans, $\mathbb{A}$ and $\mathbb{B}$:

$$\mathbb{A} = \{\{1^{Num}, dead^{A\_Words}\},$$
$$\{6^{Num}, unexploded^{A\_Words}\},$$
$$\{3^{Num}, joke^{A\_Words}\},$$
$$\{5^{Num}, spam^{A\_Words}\},$$
$$\{10^{Num}, Spanish\ Inquisition^{A\_Words}\}\}$$

and:

$$\mathbb{B} = \{\{9^{Num}, Ron\ Obvious^{B\_Words}\},$$
$$\{8^{Num}, silly\ walks^{B\_Words}\},$$
$$\{1^{Num}, parrot^{B\_Words}\},$$
$$\{3^{Num}, warfare^{B\_Words}\},$$
$$\{6^{Num}, Scotsman^{B\_Words}\}\}$$

To find the common values for the **JOIN** we proceed as follows:

$$\mathbb{I}_1 = (\mathbb{A} \circ \{Num^{Num}\}) \cap (\mathbb{B} \circ \{Num^{Num}\})$$

$$= \{\{1^{Num}\}, \{6^{Num}\}, \{3^{Num}\}, \{5^{Num}\}, \{10^{Num}\}\} \cap \{\{9^{Num}\}, \{8^{Num}\}, \{1^{Num}\}, \{3^{Num}\}, \{6^{Num}\}\}$$

$$= \{\{1^{Num}\}, \{6^{Num}\}, \{3^{Num}\}\}.$$

We can now apply a cross-superstriction to both $\mathbb{A}$ and $\mathbb{B}$ to identify the clans we require:

$$\mathbb{I}_2 = (\mathbb{A} \blacktriangleright \mathbb{I}_1)$$
$$= \{\{1^{Num}, dead^{A\_Words}\}, \{6^{Num}, unexploded^{A\_Words}\}, \{3^{Num}, joke^{A\_Words}\}\}$$
$$\mathbb{I}_3 = (\mathbb{B} \blacktriangleright \mathbb{I}_1)$$
$$= \{\{1^{Num}, parrot^{B\_Words}\}, \{3^{Num}, warfare^{B\_Words}\}, \{6^{Num}, Scotsman^{B\_Words}\}\}.$$

Now we need to define the clan of all yang-functional relations:

$$\mathbb{F}_{yang} := \{R \in \mathfrak{P}(\mathcal{G} \times \mathcal{G}): R \text{ is yang-functional}\}.$$

If you do not immediately remember what yang-functional means in respect to a relation, here's a reminder:

A relation R is yang-functional if, and only if, $a^x, b^x \in R$ implies $a = b$.

The reason we are defining this useful clan is that we want to take a cross-union that includes only certain yang-functional relations. This obliges us to define a partial cross union of two clans. The definition is as follows:

For two clans $\mathbb{A}$ and $\mathbb{B}$,

$$A \blacktriangledown_{F_{yang}} B = (A \blacktriangledown B) \cap F_{yang}.$$

Using our example to show how this works, we need to take the partial cross-union:

$$J = I_2 \blacktriangledown_{F_{yang}} I_3 = (I_2 \blacktriangledown I_3) \cap F_{yang}.$$

$$
\begin{aligned}
I_2 \blacktriangledown I_3 = \{ & \{1^{Num}, dead^{A\_Words}\} \cup \{1^{Num}, parrot^{B\_Words}\}, \\
& \{1^{Num}, dead^{A\_Words}\} \cup \{3^{Num}, warfare^{B\_Words}\}, \\
& \{1^{Num}, dead^{A\_Words}\} \cup \{6^{Num}, Scotsman^{B\_Words}\}, \\
& \{6^{Num}, unexploded^{A\_Words}\} \cup \{1^{Num}, parrot^{B\_Words}\}, \\
& \{6^{Num}, unexploded^{A\_Words}\} \cup \{3^{Num}, warfare^{B\_Words}\}, \\
& \{6^{Num}, unexploded^{A\_Words}\} \cup \{6^{Num}, Scotsman^{B\_Words}\}, \\
& \{3^{Num}, joke^{A\_Words}\} \cup \{1^{Num}, parrot^{B\_Words}\}, \\
& \{3^{Num}, joke^{A\_Words}\} \cup \{3^{Num}, warfare^{B\_Words}\}, \\
& \{3^{Num}, joke^{A\_Words}\} \cup \{6^{Num}, Scotsman^{B\_Words}\}\}
\end{aligned}
$$

$$
\begin{aligned}
= \{ & \{1^{Num}, dead^{A\_Words}, parrot^{B\_Words}\}, \\
& \{1^{Num}, dead^{A\_Words}, 3^{Num}, warfare^{B\_Words}\}, \\
& \{1^{Num}, dead^{A\_Words}, 6^{Num}, Scotsman^{B\_Words}\}, \\
& \{6^{Num}, unexploded^{A\_Words}, 1^{Num}, parrot^{B\_Words}\}, \\
& \{6^{Num}, unexploded^{A\_Words}, 3^{Num}, warfare^{B\_Words}\}, \\
& \{6^{Num}, unexploded^{A\_Words}, Scotsman^{B\_Words}\}, \\
& \{3^{Num}, joke^{A\_Words}, 1^{Num}, parrot^{B\_Words}\}, \\
& \{3^{Num}, joke^{A\_Words}, warfare^{B\_Words}\}, \\
& \{3^{Num}, joke^{A\_Words}, 6^{Num}, Scotsman^{B\_Words}\}\}.
\end{aligned}
$$

So:

$$
\begin{aligned}
J = (I_2 & \blacktriangledown I_3) \cap F_{yang} \\
= \{ & \{1^{Num}, dead^{A\_Words}, parrot^{B\_Words}\}, \\
& \{6^{Num}, unexploded^{A\_Words}, Scotsman^{B\_Words}\}, \\
& \{3^{Num}, joke^{A\_Words}, warfare^{B\_Words}\}\}.
\end{aligned}
$$

The intersection removes the relations in the clan $I_2 \blacktriangledown I_3$ that are not yang-functional. For example, $\{1^{Num}, dead^{A\_Words}, 3^{Num}, warfare^{B\_Words}\}$ has two couplets 1^{Num} and 6^{Num} that have the same yang value, but their yins, 1 and 6, are unequal. Σhgn

135

We can now provide a general algebraic expression that gives the **natural join**, which we denote by $\bowtie$, of two clans joined by a single diagonal element:

$$\mathbb{A} \bowtie \mathbb{B} := (\mathbb{A} \blacktriangleright [\mathbb{A} \circ \{\{x^x\}\}]) \blacktriangledown_{\mathbf{F}yang} (\mathbb{B} \blacktriangleright [\mathbb{B} \circ \{\{x^x\}\}])$$

where $x \in (\textbf{\textit{yang}}\,(\mathbb{A}) \cap \textbf{\textit{yang}}\,(\mathbb{B}))$ and

$$\mathbf{F}_{yang} := \{R \in \mathfrak{P}(\mathcal{G} \times \mathcal{G}): R \text{ is yang-functional}\}.$$

Incidentally, as we indicated above, it is possible to carry out a **JOIN** based on more than one diagonal element. This is more complex; so we leave it as an exercise for the utterly avid mathematical reader who delights in such exercises.

We cannot pretend that the above formula is identical to a natural join of two tables. It is an algebraic parallel. So, if you have two tables of data that you can validly represent as clans, then this formula will give you the clan representation of the **INNER JOIN** of those two tables on a single column.

Our formula compresses what we did to calculate the **INNER JOIN** of our example clans into a single expression. Let's walk through that again in words:

- We identified a common column for the **JOIN**. We did this by taking the intersection of the yang sets of $\mathbb{A}$ and $\mathbb{B}$, and selecting an element x from that set.

- The column headings of a table correspond to the yang set of the clan that represents the table. Note that there could be more than one such element in the intersection. Where that is the case, more than one **INNER JOIN** is possible. If there is no such element, no **INNER JOIN** is possible.

- Having determined x as an element of the intersection, we evaluate the compositions $\mathbb{A} \circ \{\{x^x\}\}$ and $\mathbb{B} \circ \{\{x^x\}\}$.

- We then use $\mathbb{A} \circ \{\{x^x\}\}$ and $\mathbb{B} \circ \{\{x^x\}\}$ to perform cross-superstrictions on $\mathbb{A}$ and $\mathbb{B}$ thereby removing the relations that cannot participate in the **JOIN**.

- Having done that, we can perform a yang-functional cross-union of the results of our cross-superstrictions, giving us the result we require.

The LEFT JOIN

Now that we have demonstrated the **INNER JOIN**, we can turn our attention to the other species of **JOIN** that SQL enables. Take a look at Figure 13 (page 132). The **LEFT JOIN** is (2) in the illustration.

The SQL statement is:

SELECT *[list]* **FROM** *A* **LEFT JOIN** *B*, **ON** *A.value = B.value*;

A **LEFT JOIN** on our example tables *A* and *B* could be written as follows:

SELECT * **FROM** *A* **LEFT JOIN** *B* **ON** *A.Num = B.Num*;

Table 13: LEFT JOIN result		
Num	**A_Words**	**B_Words**
1	dead	parrot
6	unexploded	Scotsman
3	joke	warfare
5	spam	NULL
10	Spanish Inquisition	NULL

The outcome would be as we illustrate in Table 13. We get the result of the **INNER JOIN** plus two additional table rows comprised of row values from table **A** and a "NULL" value where values from table **B** are inserted for the **INNER JOIN**. As you should be aware by now, we cannot mimic this completely because it is mathematical nonsense. The *NULL* is a duct tape and rubberband technique used in relational databases for making something that doesn't fit into a two dimensional table pretend that it can.

What we can do is create a clan that includes all of the valid data but excludes the *NULLs*. To do this we first calculate:

$$\mathbb{L} = \mathbb{A} \blacktriangleright ((\mathbb{A} \circ Num^{Num}) \cap (\mathbb{B} \circ Num^{Num}))$$

$$= \{\{1^{Num}, dead^{A\_Words}\},$$
$$\{6^{Num}, unexploded^{A\_Words}\},$$
$$\{3^{Num}, joke^{A\_Words}\}\}.$$

This gives us all the relations from $\mathbb{A}$ that would participate in the **INNER JOIN**. We then take the complement of $\mathbb{L}$ in $\mathbb{A}$ as follows:

$$\mathbb{A} - \mathbb{L} = \{\{5^{Num}, spam^{A\_Words}\},$$
$$\{10^{Num}, Spanish\ Inquisition^{A\_Words}\}\}.$$

We now take a union of this with the natural join:

$$(A \bowtie B) \cup (A - L).$$

So:

$$(A \bowtie B) \cup (A - L) = \{\{1^{Num}, dead^{A\_Words}, parrot^{B\_Words}\},$$
$$\{6^{Num}, unexploded^{A\_Words}, Scotsman^{B\_Words}\},$$
$$\{3^{Num}, joke^{A\_Words}, warfare^{B\_Words}\},$$
$$\{5^{Num}, spam^{A\_Words}\},$$
$$\{10^{Num}, Spanish\ Inquisition^{A\_Words}\}\}.$$

Thus, the expression for evaluating a *NULL*-free algebraic approximation for the SQL **LEFT JOIN** is:

$$(A \bowtie B) \cup (A - L),$$

where $L = A \blacktriangleright ((A \circ Num^{Num}) \cap (B \circ Num^{Num}))$.

The left outer join

The so-called "left outer join", diagram (4) in Figure 13 (page 132), is simply a **LEFT JOIN** that excludes the **INNER JOIN**. The SQL statement is as follows:

SELECT *[list]* **FROM** *A* **LEFT JOIN** *B*, **ON** *A.value* = *B.value* **WHERE** *B.value* **IS NULL;**

For our example the SQL would be:

SELECT * **FROM** *A* **LEFT JOIN** *B*, **ON** *A.Num* = *B.Num* **WHERE** *B.Num* **IS NULL;**

The outcome is:

Table 14: "Left outer join" result		
Num	**A_Words**	**B_Words**
5	spam	NULL
10	Spanish Inquisition	NULL

From a data algebra point of view, the equivalent result is:

$$A - L = \{\{5^{Num}, spam^{A\_Words}\},$$
$$\{10^{Num}, Spanish\ Inquisition^{A\_Words}\}\}.$$

We have simply removed the natural join component of the **LEFT JOIN**.

The RIGHT JOIN

It should be clear to you, dear reader, that the **RIGHT JOIN,** diagram (3) in Figure 13 (page 132), involves a judicious "switching" of table *A* with table *B*. So the SQL statement is:

SELECT *[list]* **FROM** *A* **RIGHT JOIN** *B*, **ON** *A.value* = *B.value*;

A **RIGHT JOIN** on our example tables *A* and *B* could be written as follows:

SELECT * **FROM** *A* **RIGHT JOIN** *B* **ON** *A.Num* = *B.Num*;

The result table for the **RIGHT JOIN** is:

Table 15: RIGHT JOIN result		
Num	**A_Words**	**B_Words**
1	dead	parrot
6	unexploded	Scotsman
3	joke	warfare
9	NULL	Ron Obvious
8	NULL	Silly Walks

Similarly the algebraic equivalent just involves a judicious replacement of $\mathbb{A}$ with $\mathbb{B}$. The algebraic expression for evaluating a *NULL*-free algebraic approximation for the SQL **RIGHT JOIN** is:

$$(\mathbb{A} \bowtie \mathbb{B}) \cup (\mathbb{B} - \mathbb{R}),$$

$$\text{where } \mathbb{R} = \mathbb{B} \blacktriangleright ((\mathbb{A} \circ Num^{Num}) \cap (\mathbb{B} \circ Num^{Num})).$$

$$
\begin{aligned}
(\mathbb{A} \bowtie \mathbb{B}) \cup (\mathbb{B} - \mathbb{R}) = \{ &\{1^{Num}, dead^{A\_Words}, parrot^{B\_Words}\}, \\
&\{6^{Num}, unexploded^{A\_Words}, Scotsman^{B\_Words}\}, \\
&\{3^{Num}, joke^{A\_Words}, warfare^{B\_Words}\}, \\
&\{9^{Num}, Ron\ Obvious^{B\_Words}\}, \\
&\{8^{Num}, Silly\ Walks^{B\_Words}\}\}.
\end{aligned}
$$

The right outer join

The right outer join is diagram (5) in Figure 13 (page 132), and as you should expect, also involves a switching of table *A* with table *B*. The SQL statement is:

SELECT *[list]* **FROM** *A* **RIGHT JOIN** *B*, **ON** *A.value* = *B.value* **WHERE** *A.value* **IS NULL;**

For our example the SQL would be:

SELECT * FROM *A* LEFT JOIN *B*, ON *A*.*Num* = *B*.*Num* WHERE *A*.*Num* IS NULL;

The result table for the right outer join is:

Table 16: "Right outer join" result		
Num	**A_Words**	**B_Words**
9	NULL	Ron Obvious
8	NULL	Silly Walks

From a data algebra point of view, the equivalent result is:

$$\mathbb{B} - \mathbb{R} = \{\{9^{Num}, Ron\ Obvious^{B\_Words}\},$$
$$\{8^{Num}, Silly\ Walks^{B\_Words}\}\}.$$

The FULL OUTER JOIN

The **FULL OUTER JOIN,** diagram (6) in Figure 13 (page 132), includes all the rows of both tables in one way or another. The SQL statement for it, using our example tables is:

SELECT * FROM *A* FULL OUTER JOIN *B*, ON *A*.*Num* = *B*.*Num*;

The result table is:

Table 17: FULL OUTER JOIN result		
Num	**A_Words**	**B_Words**
1	dead	parrot
6	unexploded	Scotsman
3	joke	warfare
5	spam	NULL
10	Spanish Inquisition	NULL
9	NULL	Ron Obvious
8	NULL	Silly Walks

The algebraic expression for evaluating a *NULL*-free algebraic approximation for the SQL **FULL OUTER JOIN** is:

$$(\mathbb{A} \bowtie \mathbb{B}) \cup (\mathbb{A} - \mathbb{L}) \cup (\mathbb{B} - \mathbb{R}).$$

In this instance, the resulting clan is:

$$\{\{1^{Num}, dead^{A\_Words}, parrot^{B\_Words}\},$$
$$\{6^{Num}, unexploded^{A\_Words}, Scotsman^{B\_Words}\},$$
$$\{3^{Num}, joke^{A\_Words}, warfare^{B\_Words}\},$$
$$\{5^{Num}, spam^{A\_Words}\},$$
$$\{10^{Num}, Spanish\ Inquisition^{A\_Words}\},$$
$$\{9^{Num}, Ron\ Obvious^{B\_Words}\},$$
$$\{8^{Num}, Silly\ Walks^{B\_Words}\}\}.$$

The exclusive full outer join

Finally, the exclusive full outer join, diagram (7) in Figure 13 (page 132) excludes the **INNER JOIN**. The SQL statement, using our example, is:

SELECT * FROM *A* FULL OUTER JOIN *B*, ON *A*.Num = *B*.Num WHERE *A*.Num IS NULL OR *B*.Num IS NULL;

The result table is:

Table 18: "Exclusive full outer join" result		
Num	A_Words	B_Words
5	spam	NULL
10	Spanish Inquisition	NULL
9	NULL	Ron Obvious
8	NULL	Silly Walks

The algebraic expression for evaluating a *NULL*-free algebraic approximation for the SQL so-called "exclusive full outer join" is:

$$(\mathbb{A} - \mathbb{L}) \cup (\mathbb{B} - \mathbb{R}).$$

In our example:

$$(\mathbb{A} - \mathbb{L}) \cup (\mathbb{B} - \mathbb{R}) = \{\{5^{Num}, spam^{A\_Words}\},$$
$$\{10^{Num}, Spanish\ Inquisition^{A\_Words}\},$$
$$\{9^{Num}, Ron\ Obvious^{B\_Words}\},$$
$$\{8^{Num}, Silly\ Walks^{B\_Words}\}\}.$$

From an algebraic perspective, it is not really a **JOIN** at all; it's just a union.

For the sake of those readers who appreciate illustrations with their SQL **JOIN**s, we show the full family of them in Figure 14.

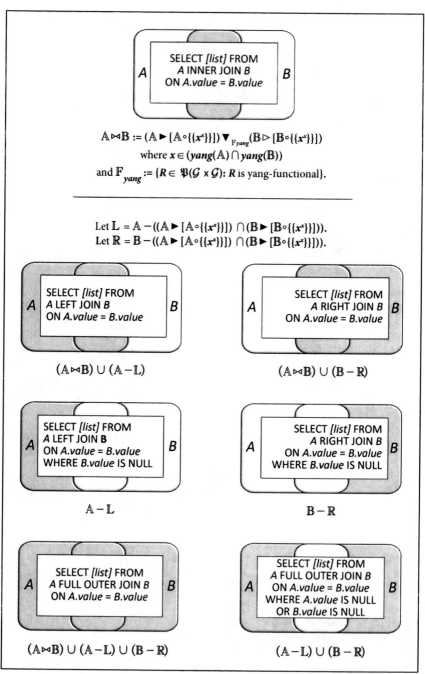

$$\mathbb{A}\bowtie\mathbb{B} := (\mathbb{A} \blacktriangleright [\mathbb{A}\circ\{\{x^x\}\}]) \blacktriangledown_{F_{yang}} (\mathbb{B} \triangleright [\mathbb{B}\circ\{\{x^x\}\}])$$

where $x \in (yang(\mathbb{A}) \cap yang(\mathbb{B}))$

and $\mathbb{F}_{yang} := \{R \in \mathfrak{P}(\mathcal{G} \times \mathcal{G}): R \text{ is yang-functional}\}.$

Let $\mathbb{L} = \mathbb{A} - ((\mathbb{A} \blacktriangleright [\mathbb{A}\circ\{\{x^x\}\}]) \cap (\mathbb{B} \blacktriangleright [\mathbb{B}\circ\{\{x^x\}\}])).$

Let $\mathbb{R} = \mathbb{B} - ((\mathbb{A} \blacktriangleright [\mathbb{A}\circ\{\{x^x\}\}]) \cap (\mathbb{B} \blacktriangleright [\mathbb{B}\circ\{\{x^x\}\}])).$

$(\mathbb{A}\bowtie\mathbb{B}) \cup (\mathbb{A} - \mathbb{L})$

$(\mathbb{A}\bowtie\mathbb{B}) \cup (\mathbb{B} - \mathbb{R})$

$\mathbb{A} - \mathbb{L}$

$\mathbb{B} - \mathbb{R}$

$(\mathbb{A}\bowtie\mathbb{B}) \cup (\mathbb{A} - \mathbb{L}) \cup (\mathbb{B} - \mathbb{R})$

$(\mathbb{A} - \mathbb{L}) \cup (\mathbb{B} - \mathbb{R})$

Figure 14: The family of JOINs with algebraic expressions

The Algebraic Playground

In this chapter, our intention has been to demonstrate that clans can represent tabular data structures and that such tables are subject to the rigors of clan algebra. While a clan is a mathematically legitimate more general data structure than a relational database table, such tables can be represented as clans of a particular kind. Algebraically, they are yang-regular clans. In Chapter 7, we demonstrated how graphs can be represented algebraically, both as relations and as clans.

If we consider tabular data and graphical data in all their possible forms (relational database, graphical database, RDF database, text database, document database, content databases, XML files, key value stores, flat files, etc.), it comprises a vast population of possible data structures and data collections. When we combine the three algebras: Couplets ($\mathcal{G}$), Relations ($\mathcal{G}$) and Clans ($\mathcal{G}$), we have a set of mathematical tools that can represent almost all of it. Why "almost"?

Hordes

Relations are sets of couplets. Clans are sets of sets of couplets. It is possible to have sets of sets of sets of couplets. Such sets are called **Hordes**. Just as the ground set for relations is $\mathfrak{P}((\mathcal{G} \times \mathcal{G})$ and the ground set for clans is $\mathfrak{P}^2((\mathcal{G} \times \mathcal{G})$, the ground set for hordes is $\mathfrak{P}^3((\mathcal{G} \times \mathcal{G})$, just the next tier of the tower.

Algebraically, databases are usually examples of a horde because they are sets of tables that can be represented as clans. The same can be said for a file system on almost any computer. The files may be relations or they may be clans, so the whole file system is likely to be a horde. Some XML files are hordes, as they can contain nested data structures within nested data structures. Additionally, some non-relational databases allow arrays (or tables) as elements of a data set. Data sets in such databases could be represented as hordes, or alternatively, the table could be represented as the yin of a couplet.

One of the beauties of data algebra is that each algebra is included within the algebra at the next level of the tower:

$$\text{Couplets}(\mathcal{G}) \xrightarrow{\text{included}} \text{Relations}(\mathcal{G}) \xrightarrow{\text{included}} \text{Clans}(\mathcal{G}) \xrightarrow{\text{included}} \text{Hordes}(\mathcal{G})$$

The level of the algebraic tower, which presides over sets of sets of sets of sets of couplets and a power set $\mathfrak{P}^4((\mathcal{G} \times \mathcal{G})$, is rarely visited because it is rarely needed. There are no restrictions on the elements of the genesis set $\mathcal{G}$. The genesis set $\mathcal{G}$ is like a sliding window. There is no restriction on its size except that it cannot be an infinite set. Its elements can even include couplets or relations or clans. This allows for a extremely varied set of sophisticated data structures without even ascending to hordes.

And if any requirement emerges that necessitates the climbing of the algebraic tower up to hordes and beyond, all the operators which are used lower down at various levels of the algebraic tower – the transposition, composition, union, intersection, complementation, superstriction, substriction, cross-union, cross-intersection, cross-superstriction and cross-substriction – can be lifted to higher levels. At those higher levels, there may be other useful operators waiting to be identified.

Taking Stock

This chapter explores the use of data algebra to provide equivalent capability to the use of the SQL **SELECT** statements and in particular, the SQL **JOIN**. The following bullet points summarize its content:

- A SQL **SELECT** that selects from a table based on column names, of the form:

  SELECT *column_name1, column_name2,...* **FROM** *table_name*;

  is achieved by a composition of the form:

  $$\mathbb{C} = \mathbb{A} \circ \{\{x^x, y^y, ...\}\},$$

  where $\mathbb{A}$ is the clan representing the table *table_name* and x, y, ... are values selected from the set *yang*($\mathbb{A}$).

- In order to create parallel expressions to various SQL statements we introduced several new operators. For Relations($\mathcal{G}$) we defined two partial operators **superstriction**, denoted by $\triangleright$ and **substriction**, denoted by $\triangleleft$, as follows:

  $$A \triangleright B := A \text{ if } A \supset B$$

  otherwise the result is undefined.

  $$A \triangleleft B := A \text{ if } A \subset B$$

  otherwise the result is undefined.

- We lifted these operators to Clans($\mathcal{G}$) to create two further operators, the **cross-superstriction,** and **cross-substriction**. These are defined as follows:

  The cross-superstriction of two clans is the clan of all the superstrictions between the relations in the original clans, and is denoted by $\blacktriangleright$.

  Also:

  The cross-substriction of two clans is the clan of all the substrictions between the relations in the original clans, and is denoted by $\blacktriangleleft$.

- A SQL **SELECT** that selects from a table based on values for particular columns, of the form:

SELECT * **FROM** *table_name* **WHERE** *column_name1 operator value*
OR *column_name2 operator value;*

is achieved by first applying a composition then a cross-superstriction.

- The general algebraic expression that gives the **natural join,** (equivalent to a SQL **INNER JOIN**) denoted by $\bowtie$, of two clans, $\mathbb{A}$ and $\mathbb{B}$, joined by a single diagonal element is:

$$\mathbb{A} \bowtie \mathbb{B} := (\mathbb{A} \blacktriangleright [\mathbb{A} \circ \{\{x^x\}\}]) \blacktriangledown_{\mathbf{F}yang} (\mathbb{B} \blacktriangleright [\mathbb{B} \circ \{\{x^x\}\}])$$

where $x \in (\textbf{\textit{yang}}\,(\mathbb{A}) \cap \textbf{\textit{yang}}\,(\mathbb{B}))$ and

$$\mathbb{F}_{yang} := \{R \in \mathfrak{P}(\mathcal{G} \times \mathcal{G})\colon R \text{ is \textbf{yang-functional}}\}.$$

- The algebraic expression for evaluating a *NULL*-free algebraic approximation for the SQL "left outer join" (the **LEFT JOIN** that excludes the **INNER JOIN** is:

$$\mathbb{A} - ((\mathbb{A} \blacktriangleright [\mathbb{A} \circ \{\{x^x\}\}]) \cap (\mathbb{B} \blacktriangleright [\mathbb{B} \circ \{\{x^x\}\}])).$$

- All other forms of SQL **JOIN** can be created by variations of the above two expressions.

- The next step up the algebraic tower from Clans($\mathcal{G}$) is to Hordes($\mathcal{G}$). A horde is a set of clans such as, for example, a database. The various operators we have discussed in this book can be lifted to that level. However, the algebras Couplets($\mathcal{G}$), Relations($\mathcal{G}$) and Clans($\mathcal{G}$) fulfill the majority of data definition and data manipulation requirements.

- If there is the need, it is possible to climb to even higher levels and lift the algebraic operators accordingly.

Chapter 9: The Potential of Data Algebra

Information theory began as a bridge from mathematics to electrical engineering and from there to computing.

~ James Gleick

COMPUTERS take data and transform it into other data. That is what they do, and it is all they do.

It may not feel like that's the fundamental process when you send an email to your mother or buy an airline ticket on the web. But it is. The computers transforming the data may be cell phones, tablets, PCs or even powerful servers hiding in air-conditioned data centers halfway around the world. They may be communicating furiously and accessing swaths of data. But no matter what the app is and how it works, what they are doing is transforming data. It is the only thing they are doing, and they are doing a great deal of it.

As you read these words, more than 2 billion computers are managing and processing upwards of 10 billion terabytes of data, and the whole digital ensemble is growing like bamboo in springtime. This voracious electronic ecosystem has swallowed up most of the information ever recorded, collected it, transformed it from one format to another and managed it. And it has spawned massive amounts of data of its own.

In its early days, the digital behemoth processed and recorded business transactions, but fairly soon it created business transactions of its own. Next it acquired an appetite for text and then became the engine for generating a great deal more. It ate its way into graphics and images and then spawned more of the same. It usurped the world of sound, consuming what existed and generating more. And finally it gobbled up video, giving birth to even more video. It consumed data, transformed data and created data.

Data algebra and the digital archipelago

The Internet re-engineered this digital world in a dramatic way. From the perspective of the average Joe and Jane, it seemed to happen almost at once. Suddenly PCs had browsers, and there were search engines and commerce sites like Amazon and eBay redefining the whole idea of retail, and a vast array of web sites to explore, from the weird to the wonderful.

Although the web grew swiftly after its birth, the average Joe and Jane had little idea of how it all came about. At first, computers were islands unto themselves; exchanging data between them called for magnetic tape. Then networking was born, but it ran at a slow pace until the advent of the PC. Of course, early PCs were also islands unto themselves, but soon there were Ethernet wires joining them to file servers and then to data center servers and mainframes.

Even so, in those dark days, digital communications between data centers was ponderous and pricey. The whole landscape changed when the magic Internet protocol stepped forward to bestow connectivity on every computer that could acquire an Internet address; digital highways sprang up and crisscrossed the globe. And, as if by natural evolution, the power of computer technology and networking technology just continued to accelerate. In almost no time at all, any person or program could, in theory, get to any data file anywhere. The realm of mobile phones and devices emerged and was assimilated, bringing us almost to the present day.

The digital archipelago of yore has evolved to become a vast digital continent with streams of data gushing into data lakes far and wide, and its evolution continues apace as it turns its attention to the embryonic Internet of Things. It is now, in this brave new era of data analytics and big data, that data algebra, a mathematical foundation for organizing, manipulating and understanding information, has been released into the digital ecosystem.

"So what?"

It is possible that many IT users and professionals will not immediately recognize the importance of this development. Most likely their education was not embroidered with colorful examples of the taming power of mathematics and the blessings it bestowed and continues to bestow on its fields of application. Even some seasoned mathematicians may not be well-versed in this.

Nevertheless, even a superficial analysis of mathematical discoveries and their impact on the world will attest to its remarkable contribution. Set aside the fact that the ancient world would never have known how to construct its temples and its monuments if its architects and engineers had not had a good grasp of geometry and a working knowledge of mechanics.

Ignore the fact that our musical system was a product of the mathematicians of antiquity, and start with a date close to the commencement of the Renaissance.

In 1494, Luca Pacioli introduced the double-entry accounting system, in a book on algebra intended for use by the schools in Northern Italy. It was a system of accounting that was picked up by the Venetian merchants of the day and has been used ever since. It provided a foundation for financial management to the mercantile economies of Europe that soon blossomed, and it stands as a foundation of the modern capitalistic economy.

In 1614, John Napier invented logarithms. At a simple practical level, they put a calculator (log tables or a slide rule) in the hands of everyone who needed to multiply numbers, particularly scientists and engineers. The mathematical contribution of logarithms was far more extensive than that, of course, finding application in many areas of applied mathematics.

Probability theory, primarily the work of Blaise Pascal and Pierre de Fermat, was far reaching, not just in its application to gaming, but in the foundation of statistics, establishing the mathematical foundation for insurance in all its variety and a plethora of industries, too.

The introduction of calculus, credited to both Newton and Liebniz, was even more profound. It is used by biologists to determine the growth rate of a bacterial culture, by physicists to estimate the rate of decay of radioactive elements and by chemists in thermodynamic calculations. It is employed by engineers of every persuasion and by architects, operations research analysts and statisticians. It is employed in almost every commercial activity, from mining to medicine, from communications to construction. The modern world would not exist without it.

"That's what?"

When mathematics is introduced into a field of activity from which it was previously absent, it can engender profound changes. The historical record demonstrates this again and again. So our expectation should be that data algebra will change the world of software in many areas, perhaps in all areas. We will consider some of the possibilities next.

Areas of Applications

There's a well-known joke which dates back to the time before every mobile phone had GPS and a mapping application. A tourist is driving through one of the more agricultural areas of England and, having become completely lost, stops and asks a local farm worker for directions to London. The local yokel thinks for a moment then scratches his chin and says: "Oooh, well, you know if I was going to London, I wouldn't start from here."

Data algebra finds itself in a similar situation to the tourist in the joke. The best time to introduce it was decades ago, about the time the relational model of data was being invented and SQL was being specified. It would have been relatively easy at that time to correct the defects of the relational model of data and the world of databases could have evolved in a less troublesome way than it did. Databases would have been more versatile and generations of developers would have acquired a better conceptual grasp of data.

But that ship sailed long ago. Now data algebra will have to be gradually retrofitted into the IT world we inhabit and which is growing at break-neck speed. The digital ecosystem currently presides over multiple zettabytes of data, scheduled to become yottabytes in the coming years. (1 YB = 10^{24} bytes). In the enormous network of billions of computers that store and connect this data, in theory, anyone with the appropriate permissions can gain access to any application or any data. Technically, this is the case, but in respect of data, the practical barriers are formidable.

The data archipelagoes

The world's data archipelagoes comprise billions of islands of data. The problem of making all the data available to any authorized user is severe for many reasons. The main reason is that most of the data that exists, including a great deal of data being created as you read this, was not created with data sharing in mind. As a consequence, most data is not easy to share.

Data algebra can and will resolve this problem in the course of time. It is capable of doing so because it provides a de facto data definition standard which subsumes all others. All data, no matter how it is stored, is already in a valid algebraic form, but most of it is not in a useful or explicit algebraic form that exposes the meaning of the data. If it were, it could be made available for use by any program that had the appropriate permissions.

The primary problem resides in the way that data is defined. All programs employ a physical map of the data (strings, floats, integers, etc.), but in most cases they do not expose the logical meaning of the data.

Data stored in files (as opposed to databases) usually includes no metadata (data which describes the meaning of the data) because the metadata is stored exclusively within the programs that use the data. If such metadata were declared for any given type of file (either posted somewhere for reference or, better, included with the file), then it would be generally available for use.

The situation with data held in databases is slightly better, since with a database there will be a database schema of some kind, providing metadata tags at some level of context. The richness of the logical data map will vary according to database type: RDBMS, RDF database, Document database, HTML database, etc. Nevertheless, it will be generally better than with data stored in files. Even in the best of circumstances, in a database environment, a good deal of the meaning of data depends on context. In the worst of circumstances, the understanding of data, whether stored in file or database, may depend on knowledge stored in the brain of the developer or the IT user. Such knowledge needs to be recorded.

Because it can span the data archipelago, data algebra could make a dramatic difference, but it will take time.

Data algebra and the OS's original sin

Several times in the history of software, various IT companies have attempted to replace an OS's file system with a database. All such efforts were at best partially successful. What was missing was an algebra of data that could precisely define all the data stored by the OS. Because of data algebra this is now feasible and, in our view, it should now be pursued by all IT businesses that preside over any OS (IBM, Microsoft, Apple, Red Hat, etc.).

In short, there needs to be an algebraic catalog and activity log. The OS will also need to implement data versioning for all data. This calls out a now-acknowledged error that was made by the IT world a long time ago.

The error lies in this fact:

a data update can be and usually is data corruption.

When you update a data item, then in most circumstances, you irretrievably destroy the record of its previous value. In doing so, you destroy the audit trail of values of the data item. The consequences of this destruction depend on how important the previous value was, and in many circumstances the consequences are not disastrous. In the early days of computing the cost of storing data was so high that the update was introduced as an awkward but convenient compromise that saved considerable cost. Traditionally databases dealt with this problem by keeping a log file of data changes. Nowadays computer resources are so inexpensive that the compromise is no longer necessary in databases or in file systems, and it would be a boon if it were done away with.

The search capability, which is an inherent part of most OSs (for example, File Explorer in Windows and Finder in OS X), is really an algebraic query capability. A far more powerful search capability could be created using a well-designed implementation of data algebra. In particular, it would provide a powerful ability for searching whole networks of computers running the same OS – assuming a secure distributed implementation of the algebraic catalog and activity log. If all OSs adopted a common data interchange standard based on data algebra, the search capability would be far more powerful still.

Databases which were implemented on any given OS that didn't internally implement the data algebra could nevertheless make their data algebraically accessible via the OS and that could be added to a data catalog. In short, the development of a truly global search capability could evolve from the practical use of data algebra within existing OSs.

Data algebra and the spreadsheet

The spreadsheet, which began life as a versatile "personal development environment," has evolved over the years by adding useful programming capabilities, by acquiring some of the functionality of a personal database, and by becoming a very useful BI tool for analyzing data and providing graphical depictions of the results. In reality, it is the dominant BI tool, although some users may not use it or think of it in that way.

To some extent, the spreadsheet's capability has been challenged in recent years by more versatile data manipulation and graphical representation tools like Tableau, Qlik, and others.

An inevitable outcome of the widespread use of data algebra will be to enhance and possibly even re-envisage the spreadsheet. What would such enhancements mean? Here is our guess:

From a utility perspective, the re-envisaged spreadsheet would be able to access any data, including data from atypical sources such as emails or PDF files. It would be augmented to enable the algebraic operations that are natural to data algebra on its data. As such, it could be used as an ETL tool as well as a BI tool. Its data representation capabilities and analytic capabilities could extend beyond the current spreadsheet. Data algebra is as at home with graphical data and hierarchical data as it is with structured data.

The Hadoop ecosystem: an OS or an archipelago?

Apache Hadoop was a new departure in the software world. Based upon an initial open source project pioneered by Yahoo!, it has given rise to a software ecosystem, consisting in part of other Apache software components and in part of complementary commercial software products. Its primary attractions are that it is built for massive scale on commodity servers, it has a rich software ecosystem and, in respect of software costs, it can be adopted relatively inexpensively.

Hadoop is gradually taking shape as an operating system of a kind that is distinct from what we usually regard as an operating system (Unix, Windows, Linux, etc.), because it is not confined to a single server. In fact it is extensible almost indefinitely. At the foundation of Hadoop is HDFS, a file system that happily distributes itself across multiple servers and is resilient. In practice, Hadoop has been implemented on many thousands of servers and in theory it could expand far beyond that. But with the relatively recent addition of Kafka, a streaming capability that can be used to connect Hadoop clusters together intelligently, it is likely that most Hadoop installations will gradually become Hadoop archipelagos.

With other components, Yarn and Mesos, Hadoop is acquiring a scheduling capability very much like an operating system and is thus evolving into an operating system for data or, as some observers have described it, the data layer OS. The major application of Hadoop has been for creating "data lakes," data storage areas that are landing points for all new data entering the organization. The concept of a data lake is evolving fast and it is rapidly being recognized as the focus point within the organization for data governance.

Data governance

Data governance can best be described as the proper regulation of corporate data. As such it means formulating and implementing rules which govern: the proper creation and maintenance of a data catalog, the registration and management of metadata, the record and maintenance of data provenance and lineage, data security, data cleansing, data distribution, and the complete management of the data life cycle from creation to final deletion or archiving.

The need for formal data governance has become increasingly apparent with the advent of the Big Data era and its inevitable expansion as it embraces the Internet of Things. Data governance is an important area for the application of data algebra. Data needs to become formally structured in respect of including information as to its origin (how, when and where it was created), its ownership and, perhaps most importantly, how and by whom it can be used.

While it is clear that data algebra will be useful within traditional operating systems, its utility within a distributed "operating system for data" will be significantly greater since it will be able to implement an extensible data catalog and it can become the engine for corporate data governance. The complexity of the data lake environment increases significantly once we take into account the increasing prevalence of data streaming and the need to build applications that process event data in real-time. Data algebra is capable of representing data in motion just as effectively as it can represent data at rest and, hence, is suited to the governance of this emerging environment.

Database query optimization

The primary purpose of a database is to provide swift access to data. Conceptually, a database is a shared pool of data that applications and database users can access concurrently. Various techniques are used to optimize the speed of retrieving data from disk. Such techniques include:

- Caching: the holding of subsets of frequently used data in local memory so that it can be accessed faster.

- Indexing: indexes provide a fast mechanism for retrieving data.

- Data structures: data is held in a convenient form that suits the query workload, such as storing data in columns.

- Data compression: compression reduces the space data items occupy, speeding up the movement of data from disk to memory or from memory to CPU.

- Query plan reuse; the caching of query plans in memory and their reuse.

These and other techniques are, to some degree, hardware agnostic. For example, replacing spinning disk with SSD does not alter the fact that SSD is slower than memory. As hardware evolves, database software adjusts to cater for the changes and might even be extensively rewritten. However, the reality remains that a computer's resources have varying costs and speeds (CPU is faster than memory which is faster than storage technology). Once data expands beyond the capacity of a single server, network speeds are factored in. Every part of the hardware configuration has limited data capacity and a maximum speed. So database software works within these parameters.

Although the techniques may vary from product to product, all database optimizers take the same general approach. Conceptually, it could be described as follows:

> There is a regularly updated collection of data that users wish to query. The pattern of queries (normally SQL queries) is analyzed in real-time as data is served up and various techniques are employed that seek to access the data as fast as possible, while balancing the available resources between multiple concurrent queries. So the optimization techniques are based upon the relationship between the stream of queries and the pool of stored data they target.

Algebraic query acceleration

Data algebra enables a different approach to query optimization, one that complements the traditional approach. Data algebra translates queries into algebraic functions and stores subsets of the query results for possible reuse. It can be thought of as a caching mechanism that focuses on query results rather than raw database data.

The underlying principle is that most query traffic is repetitive, accessing the same data, performing the same table joins and producing results

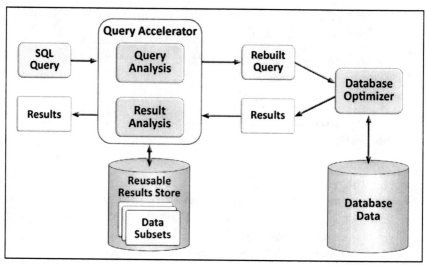

Figure 15: Query Acceleration and Database Optimization

which overlap subsets of previous query results. The algebraic accelerator stores reusable results—not all results, just those partial results that are frequently calculated. The accelerator learns more as it processes each additional query.

A diagram of how that the accelerator works with a database is shown in Figure 15. The accelerator is almost independent of the database optimizer that it complements. It intercepts all SQL traffic to the database and parses each SQL query algebraically, transforming it into an algebraic representation. It holds a store of results and subsets of results from previous queries. To determine whether there is an opportunity to reuse any of the "chunks" of stored results it consults an algebraic catalog, comparing the algebraic representation of the query with the algebraic catalog. Where some of the data it stores can be reused it reformats the SQL query to request only the data that is not held in the results store and then passes that to the database optimizer.

The database optimizer retrieves its results and passes them back to the query accelerator. It then merges those results with the cached result it is reusing and returns the complete answer to the application that issued the query.

If necessary, the accelerator could run on separate hardware to the database it complements. When implemented in that way, it will not consume any of the database's resources. It reduces the database's workload and also improves query performance. Since the accelerator has access to SQL traffic, its results stores can be adjusted for data ingest and updates.

This acceleration technique can be thought of as supplementing the database tuning role of the database administrator (DBA). Very few of the techniques that database optimizers perform automatically or that database administrators (DBAs) choose to implement duplicate the techniques of algebraic acceleration. The one exception is materialized views.

A materialized view (or snapshot) is a database object containing the results of a query. Materialized views are used, for example, to precalculate common table joins or to precompute summaries that are regularly requested. Some databases (Oracle, DB2, Microsoft SQL Server and a few others) include this feature, but most do not. When available, it is an option a DBA can manually select to improve performance.

If algebraic acceleration is implemented, the DBA no longer needs to consider using materialized views. The accelerator will identify and create effective ones automatically. In fact it will likely create many such reusable results that no DBA would identify as frequently occurring. Because of its algebraic analysis and the statistics it gathers, it can identify every candidate for reuse and store the ones that offer the greatest performance improvement. In general the most resource-intensive operations a database performs are table joins, sorts, summaries, and aggregations, and the accelerator significantly reduces this workload.

Federated queries

A federated database system (FDS) is one that can query multiple database instances as if they were a single logical database. We use the term "logical" because the participating databases remain where they physically reside on different servers or clusters, connected by a network. The components of the FDS that need to be centralized are the query capability and a metadata management capability. The metadata management component reconciles metadata definitions between the databases, enabling queries to be formulated against the whole collection of databases.

Federating several databases is often a better option than trying to merge several different databases or replicated versions of them. It eliminates the need to perform data transfers and transforms, and the federated databases remain current at all times. Trying to achieve the same capability by replicating data is usually both expensive and difficult to achieve. An FDS provides a standard query interface, allowing users to query multiple databases without needing to know anything about their physical implementation.

However, queries to federated databases are generally much slower than queries to large merged databases. An FDS has to take every query and decompose it into a set of subqueries—one for each physical database. When the results are returned, it merges them to provide the required answer. This is likely to be a slow process for two reasons. First the network adds latency to the query process. Secondly, the federated query is not optimized in a global manner—only the subqueries benefit from the optimization of the database to which they are directed.

Algebraix Data has tested its query accelerator in a relatively simple federated database environment (using two copies of MySQL). We will not describe the implementation in detail here, other than to note that the query accelerator can store the metadata details of multiple databases in algebraic form, along with the historic algebraic record it keeps of all queries.

Data virtualization

The algebraic accelerator works in a similar fashion to data virtualization software and thus it can be used in that context. Data virtualization products (for example, Cisco's Data Virtualization Platform) transfer subsets of data that are frequently accessed across a networked environment to reduce access times.

Because the algebraic accelerator can hold a full catalog of data across multiple databases and data stores, it can take on this role—replicating whole database tables or subsets of them. When doing so, it will reduce the latency of queries to individual database instances.

Global search

One of the great strengths of data algebra is its ability to completely define metadata, no matter what kind of data structure it is held in, from a simple file to a versatile database. We have already noted that data algebra could be used to define a rich file directory for all the data an operating system presides over.

Consider the situation where you want to get hold of data that's stored somewhere. If you are running Windows on a PC, you will likely use Explorer to find what you seek. In the Mac OS X environment you will use the Finder. If the data is in a relational database, you will use a SQL query to find it. If it is in a document database, then maybe you will use JSON. If it is in an XML database, then maybe you will use XQuery. For an RDF database, maybe you will use SPARQL. There is no standard way of requesting data; there are many different ways.

Data algebra has proved it can handle all these requests. It is thus possible to conceive of a heterogeneous capability to access data across different brands of database, data store, or file system. The extra software development work needed to achieve it would, among other things, require the harmonization of different SQL dialects, which is no small task. Nevertheless, once this was done, a unique and very powerful level of federated query acceleration would become available.

Finally, there is the issue of search capabilities, such as those provided by Google or Bing. In theory, such search capabilities can also be represented algebraically. This is an area that Algebraix Data has not put R&D effort into, and thus it is not yet clear whether data algebra could improve significantly on what the major search companies are doing. It is simply yet another possibility that might move the IT industry towards a global search capability.

Currently, if you need to access data there is a broad variety of databases, data stores, and file systems, and a bewildering number of different query languages and access techniques that you may need to know to get at the data. The ineffectiveness and inefficiency of data search, as a general activity, should be clear from this. If you are not convinced, you can consult various reports which, for example, suggest that knowledge workers spend 20-30% of their time searching for data.

Given the nature of the data archipelago that currently exists, this is perhaps not surprising. Even searching for data on your own PC is fraught with frustration. But if the whole archipelago came to be defined algebraically, it would gradually cease to be an archipelago. All data would be accessible algebraically. A general purpose query language for accessing data of any kind could and would be formulated.

As a consequence, this particular user frustration and inefficiency would be addressed. And if it were addressed, all of these distinctly different search capabilities might eventually be reduced to one, with variations for particular contexts. It would take time and the problem of searching the whole universal mass of data would still be technically challenging as it would require a distributed universal data directory.

Nevertheless, it is probably achievable as a complement to the global networking capability that is the foundation of the Internet.

Taking Stock

When mathematics is introduced into a field of activity from which it was previously absent, it can engender profound changes. We expect the same to occur in time with data algebra. The areas of software that have been identified as possible areas of application are:

- An algebraic data directory for an OS. Eventually, this would likely lead to the development of a universal distributed algebraic directory.

- An algebraic directory for environments like Hadoop and Spark.

- Algebraic catalogs for databases.

- Database query acceleration and optimization.

- Federated query capabilities and optimization.

- Data virtualization.

- Ultimately, a global search capability.

Chapter 10: Data Algebra And The Blockchain

You never change things by fighting the existing reality. To change something, build a new model that makes the existing model obsolete.

~ Richard Buckminster Fuller

———— ∞∞∞ ————

IN 2017, Algebraix Data Corporation reviewed its activities to determine what business initiatives the company might pursue and, in parallel, what might be done to encourage the general adoption of data algebra. It was clear that the popularization of data algebra would be very difficult to achieve via any type of database technology. The database market is now undermined by inexpensive open source alternatives: MySQL, PostgresSQL and the rise of data stores like Hadoop and Spark. If it were possible to create a competitive database technology, there would be no guarantee that it would become popular and popularize data algebra. The company was looking for a green field situation where data algebra might make an impact that could lead to its general adoption.

Two related developments appeared on Algebraix' radar screen in 2016. The first was the growing popularity of the blockchain and the second the increasing credibility of the InterPlanetary File System (IPFS) project. Together these provided the type of opportunity the company sought, where data algebra could become a foundational element of a new direction in the universe of data.

The Algebraix blockchain venture

In the early months of 2017 Algebraix considered how to get involved in the evolution of both of these technologies. Ultimately, it chose to pursue a blockchain development project that involved data storage and data management. Different possibilities were reviewed, including a blockchain-based data query service, a data market where companies could publicly sell access to some of their data, and a personal data vault where individuals could securely store their personal data.

The last of these possibilities – the personal data vault – was selected, and expanded to include a way for data vault users to monetize their data by being paid to watch advertisements. Put simply, the idea was to create a mobile phone app which connected advertisers to app users, allowing advertisers to target users by querying anonymized personal data.

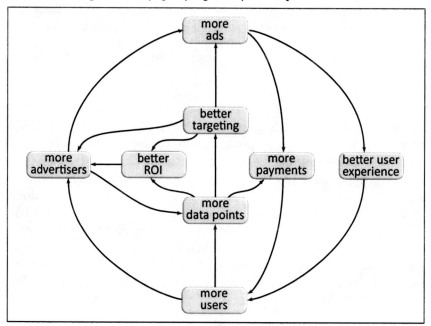

Figure 16: The Algebraix Business Model in Overview

The dynamics of the business model are illustrated in Figure 16. Starting at the bottom of the circle and following its path: More users will attract more advertisers, who will publish more ads leading more payments to users and a better user experience. These will in turn attract even more users.

However, as the diagram illustrates, there are other factors involved. Starting again at the bottom but traveling upwards: With more users there will be more data points (by volume) for advertisers to employ. Advertisers will requests new data points, which users will willingly provide, increasing the variety of data that advertisers can target. So more data points (by volume and variety) will lead to better targeting, which means better ROI for advertisers, which will attract more advertisers.

Also, better targeting will lead to more ads leading to more payments and more users. Users will accurately associate more data points with an increased opportunity to earn (more payments).

In summary the idea was for a self-reinforcing commercial environment with more users attracting more advertisers which in turn attract more users.

Of blockchains and tokens

The Algebraix approach to digital advertising has some characteristics that are novel:

- It is permission-based in the sense that none of the ads interrupt the user's activity. Viewing ads is always as the result of a choice made by the consumer.

- There are no intermediaries between the advertiser and the consumer, aside from the network that connects the two.

- The user directly earns from their data and their attention, and their data is always voluntarily given and never shared in any way without their explicit permission.

- The viewer is likely to be genuinely interested in the ad.

Because it is a closed and monitored environment, the viewer of ads is always a person and can never be a software robot, as is sometimes the case with other forms of digital advertising.

The individual user's identity is protected, and never revealed.

The business model, as described, could be built without blockchain technology and an associated cryptocurrency. However it profits from using a cryptocurrency in a very obvious way. Individual payments made by advertisers for watching ads will not necessarily be large, in fact in some contexts they might be only a few cents. It would not be economic to make such payments via traditional commercial payment channels. The business needs to processes millions of such payments every day at a very low cost, and keep a ledger of all payments.

This can be achieved using blockchain technology. A blockchain is a financial ledger, a time series of financial records with each entry denoting a monetary transfer between two accounts. It distinguishes itself from any other kind of digital ledger by virtue of the following:

- It is immutable. Once a transaction is included in a block and that block of transactions is added to the blockchain, it is immutable. The entry can never be changed.

- It is robust, partly because it is widely distributed and partly because it uses cryptography to ensure its security.

- It can be stored publicly allowing anyone who cares to to query its contents.

It is thus a secure, widely distributed database of financial transactions that is safe against any kind of hacking, even though it is exposed to public view. As blockchain technology was open source from the very beginning, it has seen a frenzy of innovation and it has been widely adopted – not just in the banking sector, but in other sectors as well.

Algebraix will use the blockchain to create and manage its ALX token. Technically, ALX is a token because it has a specific set of functions within a commercial digital ecosystem. Cryptocurrencies are referred to as coins when their primary or only function is to be money, like the dollar. The ALX token is more akin to "air miles," which are a kind of currency within the commercial ecosystem of an airline.

Blockchain structure

In practice a blockchain is a linked collection of data blocks where every block contains many transactions and also a hash pointer that points to the previous block in the chain. The hash pointer consists of a pointer to the address of the previous block and the hash value of the data inside the block. This simple hashing is what makes blockchains so remarkably reliable and, it is believed, impossible to hack.

The blockchain is not distributed in the sense that the data is divided up with different parts sent to different places. It is distributed in the sense that there are many copies of the same blockchain on many different servers. A new block cannot be added without the "consensus agreement" of all the many servers and when a new block is added it is added to every copy.

In truth, a blockchain is a very expensive database. Most databases only need to configure a few copies, perhaps just the operational database and one copy to serve as backup. So they occupy just a few servers. Depending on the consensus mechanism, a blockchain may require hundreds of servers running at the same time and occupying far more storage space.

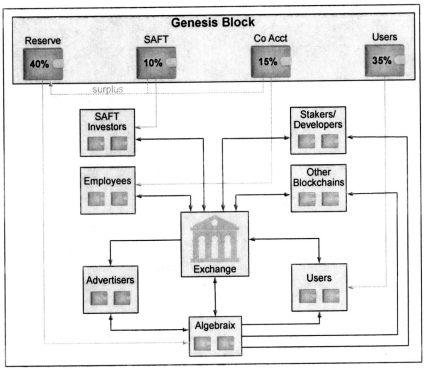

Figure 17: ALX token circulation

The circulation of ALX tokens within the Algebraix network is illustrated above. The first block in a blockchain is called the genesis block. Unless the blockchain allows the creation of new tokens (token inflation by any definition), the genesis block will contain the complete supply of token, as it does in the case of Algebraix – 100 billion tokens in total.

When the network becomes operational, those tokens begin to be distributed and new blocks are added to the blockchain. In the genesis block the ALX tokens are divided between four different corresponding wallets, containing the following proportions of the ALX token supply.:

- 35% are reserved for users of the Algebraix mobile app. These tokens will be gradually distributed to adopters of the Algebraix mobile app as rewards for downloading the application, recruiting other users and so on.

- 15%, referred to as the company account, are reserved as rewards for Algebraix staff.

- 40% form a contingency reserve, some of which will be used to reward advisers, agents, partners, external developers and possibly for block-chain integration payments.

- Finally, 10% are available for the SAFT (Simple Agreement for Future Tokens) token pre-sale. Investors who participate in the presale have tokens reserved for them which are delivered as soon as the network becomes operational.

Once the network is operational, ALX tokens will be both usable, in the sense of being able to circulate within the Algebraix network, and tradeable, on cryptocurrency trading exchanges. From then onwards, ALX tokens will gradually be distributed to users for viewing ads. They will reside within wallets within the Algebraix mobile phone application.

If there are any undistributed ALX tokens available from the SAFT or the company account, they will eventually be transferred to the contingency reserve. The idea with this and many other blockchain-based business models is that the tokens circulate usefully, enabling commercial activity, rather than sit in piles gathering dust. The initial allocation is organized in an effort to stimulate the necessary circulation.

Thus, as the diagram indicates. SAFT investors, employees, stakers, developers and other blockchains will trade their tokens over time for dollars or even other cryptocurrencies via an exchange – putting their ALX tokens into general circulation.

This trading activity is peripheral to the basic Algebraix business model, which relies on advertisers buying ALX tokens from crypto exchanges or from Algebraix to finance their advertising campaigns. The system will reward users for viewing ads and transfer ALX tokens to them accordingly. Some users will hold them, while others will choose to exchange them, putting them back into circulation for use by advertisers.

Following the money

A blockchain-based business like the one described here needs to analyze all the activity within its commercial ecosystem. In particular it will need to analyze and categorize every ALX transaction in the ecosystem.

There are two primary needs. The first is to provide a full inventory of ALX transactions to users – information they may need, for example, to

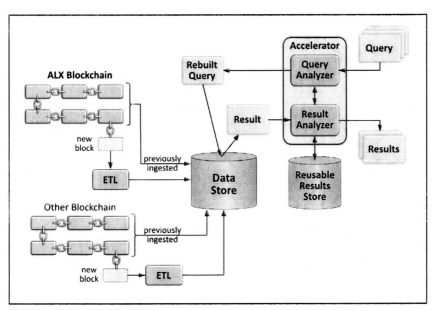

Figure 18: Fast Query Capability

complete tax returns (depending on the taxation law to which they are subject). The second is for Algebraix to be able to analyze the behavior of the population of ALX users in order to identify common behavior patterns such as: what are average earnings, what are the lowest and highest, who is holding token and who is spending it, and so on.

Algebraix could examine that data by querying the ALX blockchain, but a blockchain is not organized for fast querying, it is organized for security and immutability. As a consequence, there is a need to build a fast querying capability in preference to reading the whole blockchain for every query. This, a natural application for data algebra, is currently under development. The way it works is illustrated in Figure 18.

The diagram depicts two activities. The first is the capture of data when the ALX blockchain and perhaps one or more other blockchains create new blocks. (It is very likely that the Algebraix platform will eventually interact with and pay for services from other blockchains.)

As new transactions occur, a new block is gradually assembled and added to the ALX blockchain. As soon as this is complete, an ETL program takes the data of that block and restructures it as it loads it into the data store.

The second activity is the running of queries using a data-algebra-based results cache. New queries are analyzed to detect whether there are any prior results that can be reused. If there are, the query is rebuilt so that it excludes retrieval of the prior results. The query result is then passed to the result analyzer, if appropriate the reusable results are retrieved, and the final result is returned to the software that presented the query. This works identically to the query acceleration system described on page 156 (Figure 15).

Many of the queries that will be presented will be predictable. They will either be for individuals to create a list of transactions or they will be analytical queries, many of which will be regular and repetitive. Thus it will not only be possible to provide query acceleration but also to structure the data store in a form that is optimized for such query workloads.

There is more complexity to this application than may initially be apparent. Aside from the need for scalability – the Algebraix platform is likely to have millions of users and there will be many transactions per user per day – Algebraix plans to connect to other blockchains, perhaps many other blockchains, for the sake of the services they could provide to users of the ALX platform.

In order to map the ALX users' activity on such "partner blockchains," it will be necessary to gather data from those partner blockchains to analyze user behavior. Since data is already being gathered for this purpose, it will be possible to also include a user's independent use of these blockchains if they wish it.

Algebraix intends also to provide a multicurrency crypto wallet within its mobile application and would be able to gather details of the use of other blockchains on the users' behalf, should they request it (for tax purposes). Looking at Figure 18, it is likely that the ETL activity will be very similar for most blockchains and so would the data structure within the data store, as most blockchain transactions are simple. It thus makes sense for Algebraix to build a generic query service for all cryptocurrencies.

Because the functionality of this query service is likely to be useful to other blockchain businesses, it will be a natural candidate for open source – and it is hoped that this provokes other businesses not only to make use of the source code, but also to become familiar with data algebra.

This application will be designed to employ an algebraic query language rather than an already existing query language like SQL or SparQL. This will be the first step in defining a broadly generic query language based on data algebra.

There will also be a data store of anonymized user information that advertisers will use to target users. This too will need a data store and it too will be able to make use of algebraic query acceleration.

The InterPlanetary File System (IPFS)

Algebraix's interest in IPFS has two aspects. The first is that its blockchain system will enable users of its app to gather and securely store their personal data. Since there are already several blockchain-based storage businesses (Storj, SIA, Filecoin and others), Algebraix may use other blockchains to store the data – although it has the option of doing the storage itself. The second is that IPFS is new, ingenious, and could be made even better with data algebra.

IPFS began as an open-source project initiated by Protocol Labs in 2014. It is a peer-to-peer hypermedia protocol that is likely to become the successor to HTTP (the Hypertext Transfer Protocol), the current de facto standard for transmitting files across the Internet. HTTP has aged and will eventually need to be replaced, for several reasons.

The important reasons are: cost and scalability. In respect of current Internet costs, data providers ultimately pay the cost of data transfer. For example, when you visit a website, you connect to a computer somewhere that will serve you the information. To create the link you hop through various networks until you connect, and there is a cost to each network hop. Also the mechanism is fragile. If one link breaks the data transfer needs to be redone. With HTTP, the Internet will gradually approach the point where the cost of delivery outstrips the benefits and makes it uneconomic. This problem needs to be fixed.

The scalability problem is simply that data is not transferred in parallel (from many computers at the same time), but rather from a single computer. This is a bottleneck that can be overcome if data is reliably distributed across multiple servers that can participate in responding to a request. If this approach to data distribution reminds you of how BitTorrent works, you've got it.

IPFS has many virtues: It is a global file system with no size limit either in terms of the data stored or the number of computers connected. It uses content-addressing, allowing data to be decoupled from the source computer and stored permanently in a distributed fail-safe manner. It provides for versioning, imitating Github in its mechanism and in its capability. It addresses security problems that have plagued the HTTP-based Internet: content-addressing and content-signing protect IPFS-based sites, making DDoS attacks impossible.

You can think of IPFS as a natural complement to blockchain projects that need to store data. It enables decentralized distribution, which would make it possible, for example, to access Internet content despite a sporadic Internet service or even while offline. In effect, it organizes copies of data to be physically close to the place where they are used.

As far as Algebraix is aware, there is nothing that data algebra can do to improve the well conceived mechanisms of IPFS itself. What it can do and what Algebraix intends it to do, is to provide a layer over IPFS to enrich the metadata of the data that it stores.

It is likely that every type of data (structured data, text, images, video, etc.) will be stored and retrieved using IPFS, as will databases small and large, perhaps even up to the petabyte level.

The Adoption of Data Algebra

Arthur Schopenhauer observed that "All truth passes through three stages. First, it is ridiculed. Second, it is violently opposed. Third, it is accepted as being self-evident." Thankfully, data algebra will not have to pass through this kind of intellectual baptism. Mathematically, it is indisputable because it involves no new mathematics. It is applied set theory.

While its application of set theory is skillful and very perceptive of the nature of data, it is not in any way controversial. Mathematically, there is nothing to debate. If it is to be ridiculed and violently opposed, then such opposition will have to come from software developers and software engineers – and no doubt it will, to some extent. All previous paradigm shifts in software have met some kind of opposition.

There is even a pattern to this that can be seen in the past adoption of new ideas in software. The relational database was strongly opposed for many years before it gave way to general adoption. The same was true of object-oriented programming languages and the same was true of open source software development approach. The same will likely happen with the algebra of data.

Algebraix Data, as inventors of the algebra, will no doubt champion the mathematics as they develop their software portfolio. Elsewhere, the more mathematically versed and inquisitive software developers will become early adopters. At the same time, the paradigm will gain currency at universities, where there will be no direct resistance and in some areas of business IT, where the paradigm has particular relevance.

There is a time lag while debate rages about the importance and usefulness of the approach, and the level of take-up gradually increases from a relatively small base. Three different dynamics gradually reinforce one another in promoting acceptance of the idea:

1. More and more technology that employs the idea comes to market and is adopted.

2. Increasing numbers of graduates who think in terms of the new paradigm enter the workforce and have influence in the workplace.

3. Increasing numbers of businesses accept the idea and employ it in a strategic fashion in their implementations of software.

At some point, the old guard weakens to the point of acceptance of the new reality and – rightly or wrongly – the new paradigm dominates.

The mathematical imperative

There is another possible trajectory. With its query acceleration capability, data algebra has revealed a math-based technique that will accelerate any database or data store. Think of it like this: no matter how fast any given database is, the algebraic approach will speed it up and reduce its workload. The impact is additive and independent of any enhancements to computer hardware or the target database's approach to optimization. And data algebra can be expanded in the direction of becoming both a universal query accelerator and a global data directory.

It is likely that data algebra will lead to the invention of many other useful algebraic techniques that either simplify data manipulation or dramatically increase its speed. The last time that mathematics made such a contribution to the field of IT was fifty years ago, in 1965, with the invention of the fast Fourier transform (FFT) by Cooley and Tukey. The FFT speeds up the calculation of Fourier transforms dramatically (for an 8,000 point array it goes about 1,000 times faster, for a million point array it goes 100,000 times faster).

Fourier analysis has many areas of application, some of which are fundamental to IT. It has application in combinatorics, signal processing, imaging, probability theory, statistics, cryptography, and even option pricing in the financial markets. It is used throughout many areas of science and engineering, and in high performance computing. It is used to enable fast, large integer calculations and fast polynomial multiplications. It is used for efficient matrix vector multiplication and for efficient filtering algorithms. Its use is extensive. It was described by Gilbert Strang in 1994 as "the most important numerical algorithm of our lifetime," and perhaps it is.

But maybe it isn't. Maybe "the most important algorithm of our lifetime" is still waiting to be discovered. It may not be the case, but the possibility is there. Set theory has not been properly applied to data before now, but now it can be. Now this is possible for all data and in every context in which data is used.

If it were just set theory we were discussing here, the possibilities might not seem as promising as they are, but it is not just set theory that is enabled by the algebra of data. Set theory is the basis of all mathematics[8] and thus, once you bring set theory into the field of data, you bring all mathematics with it - at that fundamental level. What Algebraix Data Corporation has explored and worked on so far is most likely just the tip of the iceberg.

The algebra of data should have been introduced into IT a long, long time ago, and it would have been, had the original experimentation with relational "algebra" in the late 1960s and early 1970s resulted in the creation of a more universal algebra – but it didn't. As a consequence, the IT industry has been bereft of a mathematical foundation for data ever since. But now the situation is changed, and the industry can move forward.

And it will. In the words of Victor Hugo: "An invasion of armies can be resisted, but not an idea whose time has come."

✝

---

8   *Technically it includes all mathematics except for category theory. Category theory subsumes set theory.*

Appendix A: Symbols and Notation

Mathematical symbols

The table below lists all the mathematical symbols used in this book aside from the familiar algebraic symbols +, −, ×, ÷, =, ≠, <, and >.

Mathematical symbols	
Symbol	**Description**
{ }	Braces indicate sets
:=	Is defined as
∈	Belongs to
∉	Does not belong to
⊂	Subset of
⊊	Subset of and not equal to
| |	Cardinality
∘	Composition
↔	Transpose
×	Cartesian product
∅	The empty set
∪	Union
∩	Intersection
′	Complement
−	Difference
▼	Cross-union
▲	Cross-intersection
▷	Superstriction
◁	Substriction
►	Cross-superstriction
◄	Cross-substriction
⋈	Natural join

Notation standards

The table below provides a reference for all typographic standards used throughout this book. They are used to distinguish different mathematical elements and structures from each other.

Typographic standards		
Item	**Convention**	**Example**
element of a set	regular italic	a
couplet	regular italic with superscript	a^b
set	regular italic	A
set of sets	bold: font = Gauss	$\mathcal{A}$
genesis set	bold: font = Gauss	$\mathcal{G}$
power set	bold: font = Fraktur	$\mathfrak{P}$
sets of numbers	regular: font = Fermat	$\mathbb{N}$
relation	bold italic	$\boldsymbol{R}$
clan	bold: font = Fermat	$\mathbb{C}$
function	bold italic	$\boldsymbol{f}$
partition	regular italic: font = Greek	Π

Appendix B: The Algebraix Library

In parallel with Algebraix Data Corporation's decision to have a book written and published to explain and describe the algebra of data, the company also decided to have some of its developers produce a set of open source algebraic libraries in the Python language. The primary goal was to enable developers who were interested in data algebra to be able to experiment with it.

The Algebraix Technology Core Library is available as a Python library at http://algebraixlib.readthedocs.org/en/latest/. The library, algebraixlib, has been designed and built to enable developers to experiment with an algebraic approach to data and, if desired, develop software applications based on it. It embodies the fundamentals of data algebra.

Because of the constraints of code editors and the possibility of terminology confusion, there are differences between the mathematical symbols, notation and conventions found in this book and their representation in programming code. They are as follows

- A couplet is represented by "l ↦ r" in LaTeX, and the exponent notation has been dropped.

- The terms yin and yang are not used. They are replaced by "left component" and "right-component."

- Compositions are read from right to left instead of left to right, to conform with standard function notation.

- Instead of "lifting" we use the term "extension." Operators are "extended" rather than "lifted."

The library is fully documented and can be explored using the above link. What is provided below is simply a summary of the capabilities of the library.

The algebraixlib package includes subpackages as follows:

- algebraixlib.algebras package: This contains modules that represent algebras and their operations. It includes functions that return metadata and other functions that are mathematically related, but technically are not operations of the algebra. The algebras this package supports are:

  - The algebra of sets.

- The algebra of couplets.
- The algebra of relations.
- The algebra of clans.
- The algebra of multisets.
- The algebra of multiclans.

- algebraixlib.io package: This contains modules with facilities for importing and exporting data in the following formats: csv, json, mojson, rdf, xml.

- algebraixlib.mathobjects package: This package contains the modules that define the classes that represent data.

- algebraixlib.util package: This package contains utility modules with the following functions:

  - html: Present MathObjects as HTML pages.
  - latexprinter: Present MathObjects as LaTeX markup.
  - miscellaneous: Miscellaneous utility functions and classes.
  - rdf: RDF-specific data manipulation facilities.
  - test: Test utilities.

The package also includes the following submodules

- algebraixlib.extension module: These are facilities for extending operations from one algebra to another.

- algebraixlib.partition module: Operations for partitioning sets and multisets.

- algebraixlib.structure module: This module provides facilities to represent the structure of a MathObject.

- algebraixlib.undef module: These are facilities for representing and working with the concept of "undefined." Most operations are not defined for all types of data: set operations may not be defined on couplets, multiset operations may not be defined on sets and so on. When an operation is not defined for a given input, it returns the singleton Undef().

A longer term goal for this project is to develop an algebraic programming language. Data algebra provides a common language for defining all data – the "nouns and collections of nouns" that a programming language operates upon. Most of what is required for this was included in the initial Python libraries.

Data algebra also provides a range of "verbs" for applying standard algebraic transforms to data. This is the area that currently requires experimentation and further work. It will also require feedback from those developers who are willing to experiment with the language.

It may be that, in the medium term, algebraic data definitions and algebraic functions will be added to established languages that can be easily extended such as Ruby, PHP, Java, C++, C# and Objective C. There is no reason not to add such extensions to these popular languages.

Bibliography

This Bibliography includes both books and papers that may interest readers who wish to acquire a deeper understanding of data algebra and related topics.

Currently the only further information on data algebra is a paper that can be downloaded from the Algebraix Data Corporation web site. It was also used by the author in producing this book:

- Sherman, G. J. and Underbrink, J. C., "Data Algebra: Hiding in Plain Sight" (2012).[9]

The following are recommended mathematical texts:

- Halmos, P. R., *Naïve Set Theory* (Van Nostrand, 1960).

- Suppes, P., *Axiomatic Set Theory* (Dover, 1972).

- Dummitt, D. S. and Foote, R. M., *Abstract Algebra* (Wiley, 2003).

The following are papers relevant to the contents of this book:

- Kuratowski, K., "Sur la notion de l'ordre dans la théorie des ensembles," *Fundamenta Mathematicae*, 2 (1921): 129-131.

- Tarski, A. P., "On the calculus of relations," in *Journal of Symbolic Logic*, 6 (1941): 73-89.

- Codd, E. F., "A Relational Model of Data for Large Shared Data Banks," *Communications of the ACM* 13, 6 (1970): 377-387.

---

9   *The paper can be downloaded from http://www.algebraixdata.com/*

Authors' Biographical Details

Professor Gary J. Sherman

Gary Sherman's long history of studying (Ph.D. - Indiana University, 1971), teaching (Professor Emeritus, Rose-Hulman Institute of Technology, 1971-2006) and doing mathematics (29 refereed publications and Principal Mathematician, Algebraix Data Corporation, 2008-2014) is distinguished by, well, a lack of jail time.

Robin Bloor, Ph.D.

Robin Bloor, once a software developer, has spent most of his professional life as a technology analyst and IT consultant. He is also a frequent blogger and published author. He has a degree in Mathematics from Nottingham University (UK), a Ph.D. in Computer Science from Wolverhampton University (UK) and, like Professor Sherman, a surprising inexperience of the jailhouse.

Index

U

unary operator 73
union 18, 22, 27, 28, 31, 32,
33, 37, 38, 43, 55, 93, 94, 96,
98, 99, 100, 101, 117, 134, 135,
136, 138, 141, 144, 177

X

XML 17, 83, 101, 143

Y

yang 73, 74, 76, 79, 80, 86, 90,
93, 94, 95, 96, 106, 108, 112,
117, 121, 129, 134, 135, 136,
143, 145, 146, 179
yang-functional 95, 96, 134,
136, 146
yang-regular 96, 143
yang-set 90, 96
yin 73, 74, 76, 77, 79, 80, 86,
88, 90, 93, 95, 96, 106, 107,
108, 109, 112, 117, 121, 179
yin-functional 90, 95, 96
yin-regular 96
yin-set 117

CPSIA information can be obtained
at www.ICGtesting.com
Printed in the USA
FFOW01n2323160418
46286717-47761FF